MEAN TIME OF HIGH WATER AT GLASGOW, 1875.

Days	JULY MORN.	JULY EVEN.	AUGUST MORN.	AUGUST EVEN.	SEPT. MORN.	SEPT. EVEN.	OCTOBER MORN.	OCTOBER EVEN.	NOV. MORN.	NOV. EVEN.	DEC. MORN.	DEC. EVEN.	Days
1	11 10	11 42	0 42	1 10	2 3	2 21	2 8	2 23	2 40	2 55	2 55	3 14	1
2	—	0 11	1 35	1 58	2 37	2 55	2 39	2 55	3 12	3 29	3 31	3 48	2
3	0 39	1 6	2 20	2 41	3 11	3 28	3 9	3 25	3 45	4 1	4 6	4 24	3
4	1 33	1 59	3 2	3 22	3 43	3 59	3 39	3 54	4 19	4 38	4 46	5 9	4
5	2 25	2 53	3 42	4 1	4 13	4 28	4 9	4 23	4 59	5 22	5 31	5 57	5
6	3 18	3 40	4 19	4 36	4 43	4 58	4 39	5 0	5 50	6 21	6 25	6 54	6
7	4 3	4 25	4 52	5 10	5 16	5 35	5 23	5 48	6 56	7 36	7 25	7 59	7
8	4 47	5 8	5 27	5 45	5 56	6 19	6 16	6 48	8 18	9 12	8 34	9 12	8
9	5 30	5 51	6 8	6 23	6 49	7 24	7 31	8 19	9 40	10 14	9 46	10 16	9
10	6 13	6 36	6 45	7 11	8 0	8 58	9 10	9 56	10 42	11 7	10 44	11 10	10
11	6 59	7 22	7 42	8 17	9 49	10 33	10 34	11 5	11 30	11 50	11 36	—	11
12	7 48	8 17	9 0	9 45	11 9	11 38	11 29	11 49	—	0 10	—	0 27	12
13	8 49	9 23	10 26	11 2	—	0 2		0 8	0 30	0 52	0 53	1 19	13
14	9 59	10 32	11 34	—	0 25	0 45	0 26	0 45	1 15	1 36	1 44	2 10	14
15	11 2	11 30	0 3	0 29	1 4	1 22	1 4	1 22	1 57	2 20	2 36	3 2	15
16	11 57	—	0 51	1 11	1 39	1 56	1 39	1 57	2 43	3 6	3 29	3 54	16
17	0 22	0 45	1 30	1 49	2 12	2 29	2 17	2 37	3 32	3 57	4 17	4 42	17
18	1 8	1 29	2 7	2 24	2 47	3 5	2 59	3 20	4 22	4 49	5 9	5 35	18
19	1 47	2 4	2 40	2 57	3 24	3 43	3 42	4 5	5 18	5 48	5 59	6 25	19
20	2 23	2 41	3 15	3 33	4 1	4 21	4 29	4 53	6 20	6 54	6 52	7 20	20
21	3 0	3 17	3 52	4 10	4 42	5 4	5 19	5 50	7 31	8 9	7 49	8 21	21
22	3 37	3 57	4 27	4 46	5 28	5 56	6 25	7 5	8 50	9 29	8 54	9 29	22
23	4 15	4 35	5 6	5 28	6 29	7 7	7 51	8 44	10 2	10 32	10 2	10 33	23
24	4 55	5 15	5 51	6 16	7 53	8 47	9 34	10 14	10 58	11 20	11 1	11 27	24
25	5 36	5 59	6 45	7 19	9 42	10 30	10 45	11 11	11 42	—	11 51	—	25
26	6 22	6 49	7 57	8 43	11 6	11 36	11 35	11 56	0 3	0 23	0 13	0 35	26
27	7 16	7 45	9 36	10 28	—	0 3	—	0 16	0 41	0 59	0 55	1 15	27
28	8 19	9 0	11 8	11 42	0 25	0 46	0 33	0 51	1 15	1 32	1 34	1 53	28
29	9 44	10 26	—	0 12	1 4	1 21	1 8	1 23	1 48	2 5	2 10	2 29	29
30	11 4	11 38	0 39	1 3	1 37	1 52	1 38	1 54	2 21	2 38	2 46	3 4	30
31	—	0 11	1 25	1 46			2 10	2 26			3 23	3 40	31

To find the time of HIGH WATER at the following places, add to or subtract from the time given in the Table, as required:—

Place	H. M.	Place	H. M.	Place	H. M.
Aberdeen,	sub. 0 17	Dublin,	sub. 2 5	Limerick,	sub. 1 12
Ardrossan,	sub. 1 45	Dunbar,	add 0 50	London,	add 4 0
Ayr,	sub. 1 45	Dundalk,	sub. 2 21	Londonderry,	add 4 31
Ballyshannon,	add 4 0	Dundee,	add 1 15	Montrose,	add 0 10
Banff,	sub. 0 49	Falmouth,	add 3 40	Newcastle,	add 3 6
Bantry Harbour,	add 2 30	Fort-William,	add 4 2	Oban,	add 3 45
Belfast,	sub. 2 34	Galloway, Mull	sub. 2 2	Peterhead,	sub. 0 43
Berwick,	add 1 0	Galway,	add 3 18	Portpatrick,	sub. 2 5
Bristol,	add 5 56	Grangemouth,	add 1 15	Port Rush,	sub. 4 41
Campbelton,	sub. 1 45	Greenock,	sub. 1 12	Rothesay,	sub. 1 25
Cantyre, Mull of,	sub. 2 42	Havre,	sub. 3 26	Sligo,	add 4 0
Carnarvon,	sub. 3 44	Hull,	add 5 12	Stirling,	add 2 12
Cork Harbour,	add 3 44	Inveraray,	sub. 1 30	Stranraer,	add 4 50
Cromarty,	sub. 1 23	Inverness,	sub. 1 22	Tobermory,	add 3 48
Donaghadee,	sub. 2 5	Kirkcudbright,	sub. 2 7	Waterford,	add 4 50
Donegal,	add 4 0	Leith Pier,	add 1 0	Wick,	sub. 1 55
Douglas,	sub. 2 5	Liverpool,	sub. 1 54	Wicklow,	sub. 2 50

The mean duration of the ebb and flow of a tide is about 12 hours 25 min., that is, half the lunar day of 24 h. 50 min., the period elapsing between successive returns of the moon to the same meridian.

When — occurs it indicates that there is no tide. The height to which the tide rises varies every day. Spring tide—or when the tide rises to the highest—is about two days after new and full moon. A tide occupies more than 12 hours. The tide previous to the — occurs so near to 12 o'clock that the next tide is thrown over the next 12 hours into the following morning or evening. Morning commences immediately after 12 o'clock midnight, and evening immediately after 12 o'clock noon.

TIDES.

In 1875 the highest tides will be those of 9th March, 8th April, 7th May, 17th September, 16th October and 15th November.

RECEIPT AND BILL STAMPS.

RECEIPT for £2 & upwards, One Penny.

The person receiving the money pays the Stamp, which he is required to cancel by writing the date and his name ...

... bills, or drafts already stamped. Receipts for money passing between master and servant, but this will not apply to travellers who travel for several houses.

BILLS AND PROMISSORY NOTES.

avoirdupois, and 14 lbs. avoirdupois are 80 ... above 17 lbs. troy. Avoirdupois ... generally used in commerce.

...RES.—4 gills are 1 pin... and 4 quarts a gallon. ... is used for liquids and ... but the following is used for dry ... —2 gallons are a peck, 4 pecks ... 8 bushels a quarter. An imper... 5 oz., and an imperial gallon 10 lbs. avoirdupois, or 12·15 lb. troy. This is the only measure allowed now by law for liquids, and all dry goods not sold by heaped measure. A Scotch pint is 104·2 cubic inches, and weighs 55 oz. troy. In the old beer measure—which is still in use in several places —18 gallons of 282 cubic inches are a kilderkin, 2 kilderkins a barrel, and 3 kilderkins or 54 gallons a hogshead, 2 hogsheads a butt. In the wine and spirit measure 84 gallons are a puncheon, 63 gallons a hogshead, 2 hogsheads a pipe, and 2 pipes a tun.
LENGTHS.—A palm is 3 and a span 9

GERMAN AND FRENCH STANDARD MEASURES.

453·25 grammes, 1 lb. avoirdupois, English.
372·96 do. 1 lb. troy, English.
1 gramme, 15·433 grains troy, E.
1 kilogramme, 2 lbs. 3 oz. 4½ dr. or 2·2049 lbs. avoirdupois, or 2·6794 lbs. troy, E.
1 myriagramme, 22·0485 lb. avoirdupois E.
1 quintal, 100 kilogrammes, or 1·97 cwt. E.
1 millier, 1000 kilogrammes, or 1 French ton, or 19·7 cwt. E.
4000 metres, 1 league, E.
1609·31 metres, 1 mile, E.
1 metre, 39·37079 inch, or 3 feet 3½ inches, E.
1 decametre, 10 metres 10·93633 yards, E.
1 kilometre, 39370·79 inch, or 5-8 of a m. E.
1 myriametre, 6·2138 ms. or 6 ms. 376 yds. E.
1 litre, 61·02705 cubic inches. E.
1 litre, 1·760773 pint or 0·220097 gal. E.
1 decalitre, 2·20097 gals. E.
1 hectolitre, 2·7512 bushels. E.
1 hectolitre, 22·00297 gals. E.
1 kilolitre, 220·0297 gals. E.
1 stere, 35·3158 cubic feet. E.
1 decistere, 3·53158 cubic feet, E.
1 hectare 2·4736 inches, E.
1 arpen, 1·0430 acres, E.
1 are, 1076·4414 square feet, E.
A kilolitre is a little over 6 brls. E.; a bush. is 36·347664 litres; a qrt. is 1·13586 litres; a gal. 4·543458 litres; a rood 10·116775 ares; an acre 0·404671 hectares. To reduce English

... 34 yards 28 inches each way. An acre ... square yards, or 69 yards 20½ inch... way. Two acres ...

... Scotch m ... 36 square ells made a fall, 40 falls a ... 4 roods an acre. The acre was 77½ ... each way, 48 old Scotch acres are 6... lish. The English acre is in general ... America.

CUBIC or SOLIDS.—There are ... cubic inches in a cubic foot; 27 cu... are a cubic yard. 40 cubic feet of ... timber, or 50 cubic feet of hewn is ... 28 cubic feet of sand, or 18 of earth ... of clay, are deemed a ton. A ton of ... is 42 cubic feet, or 3·476 feet each wa... Paper—24 sheets are 1 quire, and 20 ... 1 ream.

cubic inches to gallons, divide by 277 ... reduce English cubic inches to litres, ... by 61·02705, and vice versa.
To reduce metres into our yards, ... ply by 1·09364 yards, and yards into ... by 0·914318. To convert kilogramm... pounds avoirdupois, multiply by ... or pounds into kilogrammes, by ... To convert litres into cubic inches, ... ply by 61·02705 ; the contrary by ... To convert hectolitres into imperial ... multiply by 2·7513; contrary, 0·363... convert hectares into acres, multi... 2·473614; acres into hectares by 0... For kilometres into miles, by 0·621... miles into kilometres, by 1·6102. F... eral calculations, the kilometre ... reckoned as 5-8ths of a mile.

These are readily converted into ... ples of the standards, by prefixing d... 10 times, by hecto for 100 times, by ... 1000 times, as kilogramme, 1000 gra... and by merely changing the decima... one place further to the right fo... two for hundreds, &c. To divide, ... deci for the 10th part, centi for the ... part, and milli for the 1000th part, as ... gramme, the 100th part of a gramm... terms for the multiplying are Greek ... for dividing are Latin.

DANGEROUS WORK

Arthur Conan Doyle, third from left, 12 July 1880.
(photograph by W.J.A. Grant, courtesy of Hull Maritime Museum.)

DANGEROUS WORK

Diary of an Arctic Adventure

———

ARTHUR CONAN DOYLE

Edited by
Jon Lellenberg & Daniel Stashower

The University of Chicago Press
CHICAGO AND LONDON

The University of Chicago Press, Chicago 60637
The University of Chicago Press Ltd., London

Published 2012.

Printed in the United States of America

21 20 19 18 17 16 15 14 13 12   2 3 4 5

ISBN-13: 978–0–226–00905–6 (cloth)
ISBN-13: 978–0–226–00886–8 (e-book)
ISBN-10: 0–226–00905–x (cloth)
ISBN-10: 0–226–00886–x (e-book)

Library of Congress CIP data
Doyle, Arthur Conan, Sir, 1859–1930, author.
  Dangerous work : the diary of an Arctic adventure / by Arthur Conan Doyle ;
edited by Jon Lellenberg and Daniel Stashower.
    pages   cm
  ISBN-13: 978-0-226-00905-6 (cloth : alkaline paper)
  ISBN-10: 0-226-00905-x (cloth : alkaline paper)
  ISBN-13: 978-0-226-00886-8 (e-book)
  ISBN-10: 0-226-00886-x (e-book) 1. Doyle, Arthur Conan, Sir, 1859–1930
—Diaries. 2. Arctic regions—Description and travel. 3. Whaling—Arctic
regions. 4. Authors, Scottish—19th century—Diaries. I. Lellenberg, Jon L.,
editor. II. Stashower, Daniel, editor. III. Title.
  PR4623.A415 2012
  823'.8—dc23

                              2012021908

  ♾ This paper meets the requirements of ANSI/NISO z39.48–1992
                  (Permanence of Paper).

Contents

Acknowledgments

———

Tʜᴇ ᴇᴅɪᴛᴏʀs are grateful for their assistance and encouragement to Christy Allen; Philip Bergem; Peter Blau; Catherine Cooke, Marylebone Library; Alison Corbett; Professor John Corbett, University of Macau; Richard Espley, National Maritime Museum, Greenwich; George Fletcher; Douglas Garden, Shetland Library, Lerwick; Michael Gunton, Laura Weston, and Dianne Cawood, Portsmouth Central Library; Stuart N. Frank, Senior Curator, New Bedford Whaling Museum; Roger Johnson; Timothy Johnson and Julia McKuras, University of Minnesota Libraries' Sherlock Holmes Collections; and Dr. Robert S. Katz. Aberdeen University Library, the British Library, London, the Newberry Library, Chicago, Illinois, and the Library of Congress, Washington D.C., also provided valuable support. Finally, the editors are grateful to the family members who, as heirs of Anna Conan Doyle, are the owners of Arthur Conan Doyle's whaling diary, for making it available for the preparation of this edition: Catherine Doyle Beggs, Georgina Doyle, Richard Doyle, and Charles Foley.

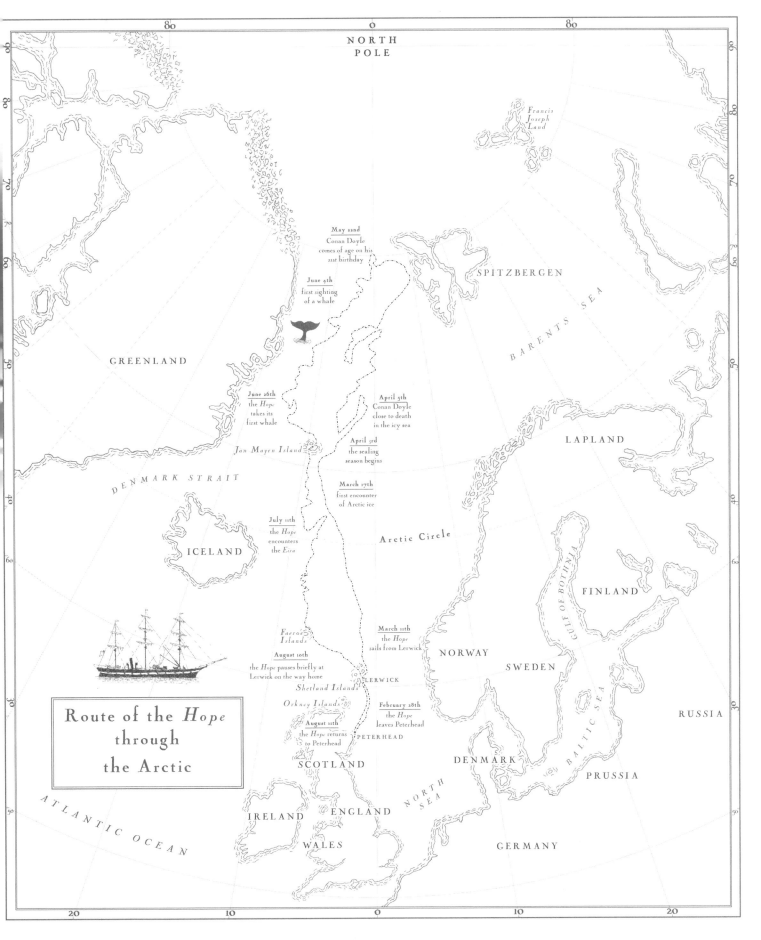

NORTH
POLE

80

0

80

90

80

70

60

*Francis
Joseph
Land*

SPITZBERGEN

70

BARENTS SEA

May 22nd
Conan Doyle
comes of age on his
21st birthday

June 4th
first sighting
of a whale

GREENLAND

June 26th
the *Hope*
takes its
first whale

April 5th
Conan Doyle
close to death
in the icy sea

LAPLAND

April 3rd
the sealing
season begins

Jan Mayen Island

March 17th
first encounter
of Arctic ice

DENMARK STRAIT

July 11th
the *Hope*
encounters
the *Eira*

Arctic Circle

ICELAND

GULF OF BOTHNIA

FINLAND

*Faeroe
Islands*

March 11th
the *Hope*
sails from Lerwick

NORWAY

SWEDEN

August 10th
the *Hope* pauses briefly at
Lerwick on the way home

Shetland Islands

LERWICK

BALTIC SEA

Orkney Islands

February 28th
the *Hope*
leaves Peterhead

RUSSIA

Route of the *Hope*
through
the Arctic

August 11th
the *Hope* returns
to Peterhead

PETERHEAD

SCOTLAND

DENMARK

PRUSSIA

ATLANTIC OCEAN

IRELAND

ENGLAND

NORTH
SEA

WALES

GERMANY

20

10

0

10

20

Arthur Conan Doyle, in practice in Southsea, Portsmouth, early 1880s.
(Courtesy of Conan Doyle Estate Ltd.)

INTRODUCTION

"I came of age at 80 degrees north latitude"

O N A MARCH AFTERNOON in 1880, a young medical student named Arthur Conan Doyle decided on a sudden impulse to suspend his studies and take a berth as ship's surgeon on an Arctic whaler. The six-month voyage took him into unknown regions, gave him unimagined sights and experiences, and plunged him into dangerous and bloody work on the ice floes of the Arctic seas. He worked harder under more difficult circumstances than he ever had before, he argued philosophy and religion with his shipmates, and he dodged death on more than one occasion. It proved to be, he said, "the first real outstanding adventure of my life."

"It came about in this way," he explained years later in his autobiography, *Memories and Adventures*:

> One raw afternoon in Edinburgh, whilst I was sitting reading hard for one of those examinations which blight the life of a medical student, there entered to me one Currie, a fellow-student with whom I had some slight acquaintance. The monstrous question which he asked drove all thought of my studies out of my head.
>
> "Would you care," said he, "to start next week for a whaling cruise? You'll be surgeon, two pound ten a month and three shillings a ton oil money."
>
> "How do you know I'll get the berth?" was my natural question.
>
> "Because I have it myself. I find at this last moment that I can't go, and I want to get a man to take my place."
>
> "How about an Arctic kit?"
>
> "You can have mine."
>
> In an instant the thing was settled, and within a few minutes the current of my life had been deflected into a new channel.[1]

Conan Doyle was only twenty at the time, and in his third year of medical studies at Edinburgh University. "Speaking generally of my university career," he would

---

1. "Whaling in the Arctic Ocean," *Memories and Adventures* (London: Hodder & Stoughton, 1924), ch. 4, quoted subsequently in this introduction without attribution.

recall, "I was always one of the ruck, neither lingering nor gaining – a 60 percent man at examinations." His typically self-effacing comment made light of a good deal of effort and accomplishment in the face of difficult circumstances. In later years he would declare with characteristic cheer that he had been raised in "the hardy and bracing atmosphere of poverty," but the remark glossed over considerable domestic turmoil and hardship, with the Doyle family changing addresses at least five times before Arthur was ten. Though it was a genteel poverty, his father Charles Doyle suffered for years from illness and alcohol, until the income from his surveyor's post ceased when he was only forty-four.

Somehow money had been found to provide young Arthur with a first-class education at Stonyhurst, a distinguished Jesuit boarding school in England, and upon graduating he felt the need to assume some of his father's responsibilities and contribute to the welfare of the large family. "Perhaps it was good for me that the times were hard," he wrote, "for I was wild, full-blooded, and a trifle reckless, but the situation called for energy and application so that one was bound to try to meet it. My mother had been so splendid that we could not fail her. It had been determined that I should be a doctor, chiefly, I think, because Edinburgh was so famous a centre for medical learning."

By now the first seeds of the Sherlock Holmes stories were sown. As a boy Conan Doyle had discovered Edgar Allan Poe, "the supreme original short story writer of all time," and would occasionally "petrify our small family circle" by reading his tales aloud. At Edinburgh University he had the good fortune to serve as an assistant to Dr. Joseph Bell, a physician whose powers of observation and diagnosis were spellbinding. At a glance Bell could often discern not only the nature of a patient's ailment, but also numerous details of his background and occupation. "To an audience of Watsons," Conan Doyle joked in later years, "it all seemed very miraculous until it was explained, and then it became simple enough." The future creator of Sherlock Holmes had already published one mystery story by the age of twenty, thrilled to receive three guineas for it when often he went without lunch in order to spend two pence upon a used book.

Conan Doyle's decision to sign onto the Arctic whaling expedition, spontaneous and reckless though it undoubtedly seemed to his industrious and thrifty mother, afforded him a rare set of opportunities. He would indulge his budding taste for adventure, and be paid for doing so. At the same time, his six months aboard ship would give him a chance to nurture his growing ambitions as a writer. Before departing for Peterhead, the Scottish port where he would join the whaler *Hope*, he augmented Claud Currie's seaman's kit with several books of poetry, philosophy,

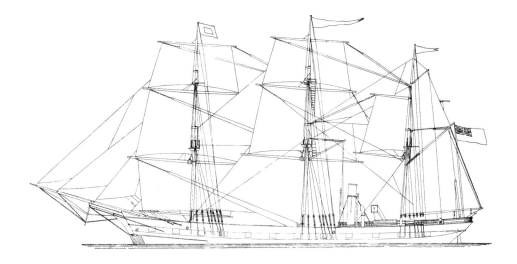

Diagram of the S.S. *Hope*, built by Alexander Hall & Co., Aberdeen, "at the height of their fame as perhaps the most successful builders of clipper ships in the world." (From Basil Lubbock's *The Arctic Whalers*, Glasgow, 1937.)

and literature, as well as blank journals in which to record his impressions of the voyage.[2] They would become a deeply personal chronicle of a young man testing himself as never before.

It is regrettable that he did not begin recording his impressions until the very moment the *Hope* sailed from Peterhead, as one would like to know also about the first eventful days. It would likely be entertaining to have a record of how Conan Doyle's hard-working mother reacted to the idea of her twenty-year-old son suspending his medical studies to go off on a hazardous junket. "Few people have any idea of the dangers incurred by whalers in the Arctic Seas," wrote the naturalist Francis Buckland in 1876: "The work they have to do is very perilous." Their ships occasionally had even to act as battering rams, he went on, "crashing a passage" through the ice by sheer force, for otherwise the ice would "impound them in a fearful prison, and subject them to all the horrors of an arctic winter."[3]

It would also be interesting to have Conan Doyle's first impressions of the *Hope*'s captain and crew, and of the Arctic whaler that would be his home for the next

2. Conan Doyle's log consists of two notebooks containing approximately 25,000 words plus seventy captioned pen and ink drawings, some hand-coloured, covering the voyage of the *Hope* from February 28 to August 11, 1880. We have annotated a transcript of it for clarity and context, adding two letters home he wrote during the experience, the first from Lerwick, capital of the Shetland Islands, the second while at sea and dispatched on a passing ship.
3. Francis Trevelyan Buckland, *Log-Book of a Fisherman and Zoologist* (London: Chapman & Hall, 1876), p. 290.

seven months. The *Hope* had been built by Alexander Hall & Co. of Aberdeen in 1873, 45 feet 5 inches in length, 28 feet 1 inch in breadth, and 17 feet in depth. In 1882 it was chosen for a dangerous Arctic rescue mission as "in all respects suitable for the work of the expedition. Strongly built, double-planked around the water-line, fortified within with iron frames, and shod with iron at the bow, she had a reputation even amongst whalers as being a ship of no ordinary capacities for encountering heavy ice; and those who sailed in her were fully persuaded that she was as good a ship for the purpose as could be procured."[4]

It carried a crew of fifty-six, Conan Doyle mentions. Its crew list has not survived, but a comparable whaler, the *Arctic* out of Dundee, carried fifty-eight: its captain, the first and second mates (also serving as harpooners), a surgeon, a steward, a first engineer, a second engineer and blacksmith, three firemen, a carpenter and a carpenter's mate, a "specksioneer" to direct the cutting up of the blubber (also serving as a harpooner), two "fast" or master harpooners and two "loose harpooners" learning the trade, a cooper who also served as a harpooner, eight line-managers (whose skill at line-coiling could be life or death), six boat-steerers for the longboats, a boatswain, a "skeeman" to direct the storage of the blubber in the ship's tanks (both serving as boat-steerers also), a ship-keeper, a cook and a cook's mate, ten Able Bodied Seamen, five Ordinary Seamen, and three cabin-boys.[5] The *Hope*'s crew was roughly divided two to one between Peterhead men and Shetland Islanders.

The *Hope* had been built to order for Captain John Gray of Peterhead, who was fifty years old at the time of this voyage. "I see him now," Conan Doyle remembered years afterwards, "his ruddy face, his grizzled hair and beard, his very light blue eyes always looking into far spaces, and his erect muscular figure. Taciturn, sardonic, stern on occasion, but always a good just man at bottom." John Gray and his older and younger brothers David and Alexander were scions of a Peterhead whaling family stretching back three generations. The "undefeatable Grays" anchored an industry which had reached its peak in the middle decades of the century, and was now tapering off amid concerns over diminishing whale populations. The Grays responded to the challenge with both forward-thinking pursuit of conservation measures, and equipping their ships with steam engines in addition to sail, in order to push farther into Arctic waters. By 1880 the industry

4. "Experiences of a Naval Officer in Search of the *Eira*," *Blackwood's Magazine*, November 1882, p. 599.
5. Albert Hastings Markham, *A Whaling Cruise to Baffin's Bay and the Gulf of Boothnia* (London: Sampson Low, 1874), p. 11.

"Fine honest fellows the men are and such a strapping lot." S.S. *Hope* crew members at the Boilyards on Keith Inch, Peterhead, in 1880. (© Aberdeenshire Heritage. Licensor www.scran. ac.uk.)

had entered its years of decline, but Peterhead whaling, says an historian of the community, "persisted longer than it might otherwise have done because of the tenacity of the Grays. It is probably also true that whaling finally died not because of the industry ailing but because the Grays were failing physically."[6] Both John and David Gray would sell their ships and retire in 1891.

This dismal end was still ahead as Gray and his crew prepared for their voyage in 1880. Conan Doyle immediately warmed to the captain, who may not have known until their first meeting that his new surgeon was not the man he expected. "I speedily found," Conan Doyle would write, "that the chief duty of the surgeon was to be the companion of the captain, who is cut off by the etiquette of the trade

6. Alexander R. Buchan, *The Peterhead Whaling Trade* (Peterhead: Buchan Field Club, 1999), p. 52.

Captain John Gray (1830–1892),
master of the *Hope*. "A really splendid man,
a grand seaman and a serious-minded Scot."
(Courtesy of Conan Doyle Estate Ltd.)

from anything but very brief and technical talks with his other officers. I should have found it intolerable if the captain had been a bad fellow, but John Gray of the *Hope* was a really splendid man, a grand seaman and a serious-minded Scot, so that he and I formed a companionship which was never marred during our long *tête-a-tête*."

Though his time in Peterhead itself was brief, Conan Doyle grasped the degree to which the town's hopes rested on the success of her diminished whaling fleet. An article in Edinburgh's newspaper *The Scotsman* in 1902 offered a wistful look back at a time of "glory and of a great industry."

There are those living in the town today who can well remember the time when no fewer than 31 vessels left its harbours to engage in the fishing, and when many of its merchants drew much wealth from the previous cargoes with which they frequently returned. Those were great days, as one can readily imagine, for the Peterheadians. There were few of them in the forties and fifties of last century who were not connected in some way or another with the industry, whether as owners of the whaling craft, members of the crews, or engaged in the boilyards, where the "blubber" was manufactured into a saleable commodity. When the spring months drew on the townspeople were wont to turn out *en masse* to give "God speed" to the fleet as it left on the voyage northward; news of the progress of the fishing was eagerly waited for towards the close of the summer, and when at last the storm-beaten craft returned to the port with hulls deep in the water, as they often did in those days, there was great rejoicing and jubilation.[7]

One study of the British whaling industry has calculated that during its three centuries of existence, at least 6,000 voyages out of thirty-five ports went to the Arctic, to either the "Greenland ground" – the waters between the eastern coast of Greenland and Norway, out to Spitzbergen – or the Davis Strait west of Greenland, including Hudson and Baffin Bays. The *Hope* worked the Greenland ground, first the harp seal grounds off Jan Mayen Island north of Iceland, and then farther north to the whaling grounds. "The main target species was the 'Greenland whale,' 'Greenland right whale' or 'bowhead,' *Baeaena mysticetus.* ... Also hunted were belugas *Delphinapterus leucas*, narwhals *Monodon monoceros*, northern bottlenose whales *Hyperoodon ampullatus*, walruses *Odobenus rosmarus*, and several species of seals."[8] Whalers spent six to seven months at sea each year, and from boyhood until retirement or death at sea never knew a normal summer below the Arctic Circle.

Standing on the *Hope*'s quarterdeck as the fleet put to sea in 1880, Conan Doyle was swept up by the pageantry and tradition. "It was, I find by my log, on February 28 at 2 p.m. that we sailed from Peterhead, amid a great crowd and uproar," he would recall.[9] The realities of life at sea made themselves felt soon enough. As the *Hope* headed north for Lerwick, the principal Shetlands port, it sailed into foul weather and menacing winds. "We just got into Lerwick Harbour before the full force of the hurricane broke," Conan Doyle recalled in his autobiography, "so great that lying at anchor with bare poles and partly screened we were blown over to an acute angle. If it had taken us a few hours earlier we should certainly have lost our boats – and the boats are the life of a whaler."

7. "The Peterhead Whalers," *The Scotsman* (Edinburgh), November 19, 1902.
8. See Sidney Brown, Arthur Credlund, Ann Savours and Bernard Stonehouse, "British Arctic whaling logbooks and journals: a provisional listing," in *Polar Record*, January 2008, pp. 311–12.
9. In fact Conan Doyle could have cited either his personal diary, which he called his log, or the official log of the vessel, which he also kept. A whaler's surgeon "frequently was required by the master to see that the ship's log was kept up to date," says Buchan's *Peterhead Whaling Trade*, p. 56, and so it was for Conan Doyle. It has vanished since 1937 when Basil Lubbock's *The Arctic Whalers* (Glasgow: Brown, Son & Ferguson Ltd.) said it was "in Conan Doyle's neat handwriting" and quoted the August 4th entry, one considerably more matter-of-fact than Conan Doyle's entry in his own diary that day, as we shall see.

Lerwick, the capital of the Shetland Islands, seen from the harbour: "A dirty little town with very hospitable simple inhabitants." (Courtesy of Shetland Museum.)

It would be more than a week before the weather calmed sufficiently. During that time Conan Doyle came to appreciate the Shetlanders' isolation on their remote archipelago. "I spoke to one old man there who asked me the news," he recorded. "I said, 'The Tay bridge is down,' which was then a fairly stale item. He said, 'Eh, have they built a brig over the Tay?'" Conan Doyle did not care much for Lerwick, but the eleven days he spent there in early March saw the town at its most bustling and boisterous during the now-gone era of the Scottish whaling fleets. In 1923, an historian of the town asked:

> How many can recall the days when in the early days of February and March the harbour was gay with a fleet of Greenland whalers, many of them fine-looking craft – brigs, barques, barquentines, mostly three-masted; steamers and sailing ships, some of them painted with portholes, to resemble ships of war; their flags fluttering in the breeze; their very-much-alive crews ashore, where they speedily got much more alive for a time, and afterwards half dead; when the town was filled with men from the country seeking to be taken on for the voyage; when the shipping offices were besieged day and night, packed with men who having signed on were being supplied with the many things needed for the Arctic voyage?[10]

10. Thomas Manson, *Lerwick During the Last Half Century* (Lerwick: T. & J. Manson, *Shetland News* Offices, 1923), p. 141.

The *Hope* in Lerwick's harbour: "We shifted our berth the other day and now lie apart from the other ships with the *Windward*." (Courtesy of Shetland Museum.)

By March 11th the foul weather had abated, and as the *Hope* departed Lerwick and continued north, Conan Doyle's work as surgeon began in earnest. "As my knowledge of medicine was that of an average third year's student," he would write, "I have often thought that it was as well that there was no very serious call upon my services."[11] This light-hearted recollection of later years glossed over a sad milestone in his medical career, however: the first death of a patient of his, from a serious intestinal disorder the young medico was powerless to treat effectively at sea.

Conan Doyle had already tended to injuries he himself had caused. "In my student days boxing was a favourite amusement of mine," he recalled, "for I had found that when reading hard one can compress more exercise into a short time in this way than in any other. Among my belongings, therefore, were two pairs of battered and discoloured gloves. Now it chanced that the steward was a bit of a fighting man; so when my unpacking was finished he, of his own accord, picked up the gloves and proposed that we should then and there have a bout."

The steward was a powerful fireplug of a man who became one of Conan Doyle's favourites among the crew. "I can see him now," he wrote later, "blue-eyed,

---

11. "Life on a Greenland Whaler," *The Strand Magazine*, January 1897 (reprinted in this volume).

yellow-bearded, short, but deep-chested, with the bandy legs of a very muscular man." Conan Doyle had no sooner pulled on his gloves than the steward pressed forward with fists raised. "Our contest was an unfair one," he reported, "for he was several inches shorter in the reach than I, and knew nothing about sparring, although I have no doubt he was a formidable person in a street row. I kept propping him off as he rushed at me, and, at last, finding that he was determined to bore his way in, I had to hit him out with some severity."

But then:

> An hour or so afterwards, as I sat reading in the saloon, there was a murmur in the mate's berth, which was next door, and suddenly I heard the steward say, in loud tones of conviction: "So help me, Colin, he's the best surrr-geon we've had! He's blackened my e'e!"

That giving the steward a black eye made Conan Doyle the best surgeon the *Hope* had had "struck me as a singular test of my medical ability, but I daresay it did no harm." (Conan Doyle had initially wondered about the red-bearded giant Colin McLean. It seemed odd that the captain had filled the first mate's position with "a little, decrepit, broken fellow, absolutely incapable of performing the duties," while McLean was signed aboard as cook's assistant. As soon as the ship cleared the harbour, however, the two men swapped places: the brawny McLean took up the duties of first mate, while his spindly crewmate disappeared into the galley. McLean was illiterate, and not eligible for a first mate's certificate.)

Less than a week after sailing from Lerwick, the *Hope* reached the open icefields. "What surprised me most in the Arctic regions," Conan Doyle reported, "was the rapidity with which you reach them. I had never realized that they lie at our very doors." The date, he recorded in his diary, was March 17. "I awoke one morning to hear the bump, bump of the floating pieces against the side of the ship," he recalled, "and I went on deck to see the whole sea covered with them to the horizon. They were none of them large, but they lay so thick that a man might travel far by springing from one to the other. Their dazzling whiteness made the sea seem bluer by contrast, and with a blue sky above, and that glorious Arctic air in one's nostrils, it was a morning to remember."

An agreement between Britain and Norway prohibited seal-hunting until after the March breeding season, so the crew used the time to track schools of seals to the main pack. "When you do come upon it, it is a wonderful sight," Conan Doyle wrote. "From the crow's nest at the top of the main-mast, one can see no end of them. On the furthest visible ice one can still see that sprinkling of pepper

An earlier whaling vessel amid the "heaviest of rafter and hummock ice" in 1869. (From William Bradford, John L. Dunmore and George Critcherson, *The Arctic Regions*, London: Sampson Low, 1973.)

grains." Once again, though, a run of bad weather interfered with their plan. "The *Hope* was one of the first to find the seal-pack that year," Conan Doyle said, "but before the day came when hunting was allowed, we had a succession of strong gales, followed by a severe roll, which tilted the floating ice and launched the young seals prematurely into the water. And so, when the law at last allowed us to begin work, Nature had left us with very little work to do."

Even so, on April 3rd, the crew fanned out across the ice with clubs in hand. Conan Doyle was determined to serve in the sealing party as well as surgeon, but as he started to join the hunt Captain Gray ordered him back – regarding the ice as too dangerous for a novice. "My remonstrances were useless," he reported, "and, at last, in the blackest of tempers, I seated myself upon the top of the bulwarks, with my feet dangling over the outer side, and there I nursed my wrath, swinging up and down with the roll of the ship. It chanced, however, that I was really seated upon a thin sheet of ice which had formed upon the wood, and so when the swell threw her over to a particularly acute angle, I shot off and vanished into the sea between two ice-blocks." Conan Doyle scrambled back on board where he found his misfortune had an unexpected benefit: "the captain remarked that as I was bound to fall into the ocean in any case, I might just as well be on the ice as on the ship."

Inexperienced on the ice, Conan Doyle fell in twice more and finished the day in bed while his clothes dried out in the engine room. "I had to answer to the name of 'the great northern diver' for a long time thereafter," he laughed.

Despite the inauspicious start, Conan Doyle became more comfortable on the ice, and went off hunting by himself as other crew members did. One afternoon, while crouching over a freshly-killed seal, he slipped backwards off the edge of the ice into the sea. "I had wandered away from the others," he said, "and no one saw my misfortune. The face of the ice was so even that I had no purchase by which to pull myself up, and my body was rapidly becoming numb in the freezing water." Death would occur in just a few minutes if he could not get back onto the ice. For several moments he clawed at the smooth edge until at last, with a desperate effort, he managed to grab the hind flipper of the seal he had been skinning. For a few tense moments there was "a kind of nightmare tug-of-war," with Conan Doyle trying to ease himself up onto the ice before the animal came sliding off upon him. At last he got one knee out of the water and flopped onto the ice beside the dead creature. Once again the "great northern diver" finished the day in bed: "my clothes were as hard as a suit of armour by the time I reached the ship, and I had to thaw my crackling garments before I could change them."

Conan Doyle later admitted to qualms over seal hunting. "It is brutal work," he wrote, "though not more brutal than that which goes on to supply every dinner-table in the country. And yet those glaring crimson pools upon the dazzling white of the icefields, under the peaceful silence of a blue Arctic sky, did seem a horrible intrusion. But an inexorable demand creates an inexorable supply, and the seals, by their death, help to give a living to the long line of seamen, dockers, tanners, curers, triers, chandlers, leather merchants, and oil-sellers, who stand between this annual butchery on the one hand, and the exquisite, with his soft leather boots, or the savant, using a delicate oil for his philosophical instruments, upon the other."

In June, the *Hope* resumed northwards and the whale hunt began. "It is seldom that one meets anyone who understands the value of a Greenland whale," Conan Doyle later wrote: "A well-boned and large one as she floats is worth today something between two and three thousand pounds. This huge price is due to the value of whalebone, which is a very rare commodity, and yet is absolutely essential for some trade purposes." As he had a stake in the profits, he took a natural interest in the hunt's success. The *Hope* carried eight whaling boats, but it was usual to send out only seven while the so-called "idlers" – those whose duties were different – remained aboard. On this voyage, perhaps at Conan Doyle's instigation, the idlers volunteered to man the extra boat. They soon made it, in his estimation, one of the most efficient. "We were all young and strong and keen,"[12] he later told an interviewer, "and I think our boat was as good as any."[13]

"It is exciting work pulling on to a whale," he would write:

> Your own back is turned to him, and all you know about him is what you read upon the face of the boat-steerer. He is staring out over your head, watching the creature as it swims slowly through the water, raising his hand now and again as a signal to stop rowing when he sees that the eye is coming round, and then resuming the stealthy approach when the whale is end on. There are so many floating pieces of ice, that as long as the oars are quiet the boat alone will not cause the creature to dive. So you creep slowly up, and at last you are so near that the boat-steerer knows that you can get there before the creature has time to dive – for it takes some little time to get that huge body into motion. You see a sudden gleam in his eyes, and a flush in his cheeks, and it's "Give way, boys! Give way, all! Hard!" Click goes the trigger of the big harpoon gun, and the foam flies from your oars. Six strokes, perhaps, and then with a dull greasy squelch the bows run upon something soft, and you and your oars are sent flying in every direction. But little you care for that, for as you touched the whale you heard the crash of the gun, and know that the harpoon has been fired point-blank into the huge, lead-coloured curve of its side. The creature sinks like a stone, the bows of the boat splash down in the water again, and there is the line whizzing swiftly under the seats and over the bows between your outstretched feet.
>
> And this is the great element of danger – for it is rarely indeed that the whale has spirit enough to turn upon its enemies. The line is very carefully coiled by a special man named the line-coiler, and it is warranted not to kink. If it should happen to do so, however, and if the loop catches the limbs of any one of the boat's crew, that man goes to his death so rapidly that his comrades hardly know that he has gone. It is a waste of fish to cut the line, for the victim is already hundreds of fathoms deep.

"Hold your hand, mon," cried the harpooner, as a seaman raised his knife on such an occasion. "The fish will be a fine thing for the widdey." It sounds callous, but there was philosophy at the base of it.

---

12. Forgetting that, as recorded in his log, one of the "mob boat" was the oldest of the *Hope*'s crew.
13. The interviewer, in the *New York World*, July 28, 1907, was Anglo-Irish theatre manager and novelist Bram Stoker, whose *Dracula* had been published ten years earlier. The interview, "Sir Arthur Conan Doyle Tells of his Career and Work, his Sentiments towards America, and his Approaching Marriage," is in *Sir Arthur Conan Doyle: Interviews and Recollections* edited by Harold Orel (London: Macmillan, 1991). In ch. 6 of *Dracula*, the Count's arrival in Whitby (an early centre of English whaling), aboard a Russian vessel from the Black Sea, is heralded by the tales of Arctic whaling an ancient sailor man named Swales tells to Mina Murray and Lucy Westenra.

Conan Doyle became a seasoned whaler, so much so that Captain Gray invited him to return for next year's voyage, as harpooner as well as surgeon. "It is well I refused," he said, "for the life is dangerously fascinating." Even so, the long months at sea fuelled a taste for adventure that remained part of his character. "To play a salmon is a royal game," he would remark, "but when your fish weighs more than a suburban villa, and is worth a clear two thousand pounds; when, too, your line is a thumb's thickness of manila rope with fifty strands, every strand tested for thirty-six pounds, it dwarfs all other experiences."

As with the slaughter of the seals, however, Conan Doyle came to have qualms about the "murderous harvest" he witnessed in the whaling grounds: "amid all the excitement – and no one who has not held an oar in such a scene can tell how exciting it is – one's sympathies lie with the poor hunted creature. The whale has a small eye, little larger than that of a bullock; but I cannot easily forget the mute expostulation which I read in one, as it dimmed over in death within hand's touch of me."

On one occasion, Conan Doyle recorded, a whale very nearly did get its revenge. He found himself part of a crew lancing a whale that had already been harpooned and played. As the rowers attempted to take up a safe position away from the thrashing tail, the wounded creature's enormous "side-flapper" rose out of the water and poised over their boat. "One flap would have sent us to the bottom of the sea," Conan Doyle recalled. "I can never forget how, as we pushed our way from under, each of us held one hand up to stave off that great, threatening fin – as if any strength of ours could have availed if the whale had meant it to descend."

The *Hope* turned homeward in early August, carrying a "scanty" cargo of two whales, some 3600 seals, and a vast assortment of polar bears, narwhals and Arctic birds. For the crew, this fairly modest haul would mark the voyage as uneventful, but Conan Doyle judged its success by a different measure. "I went on board the whaler a big, straggling youth," he would recall in his memoirs, but "I came off it a powerful, well-grown man. I have no doubt that my physical health during my whole life has been affected by that splendid air, and that the inexhaustible store of energy which I have enjoyed is to some extent drawn from the same source."

Conan Doyle also never forgot the sense of isolation he and the crew felt at having been cut off from civilization for so many long months. "We had left in exciting times," he recalled, not least because war with Russia appeared imminent. The isolation made itself felt in other ways as well. Conan Doyle recalled, as the *Hope* rounded the north of Scotland, catching sight of a woman emerging from a lighthouse on the shore, and the excited murmur of "It's a wumman!" that ran through the crew. "To appreciate a woman one has to be out of sight of one for six months," he observed. "Everyone was staring. She was well over fifty, short skirts and sea boots – but she was a 'wumman.' 'Anything in a mutch!' the sailors used to say, and I was of the same way of thinking."

Conan Doyle bade farewell to his shipmates in Peterhead and returned to Edinburgh, where his reunion with his mother was all the happier for his pay and share of the oil money, adding some fifty pounds sterling to the family purse. "Now I had a straight run in to my final examination," he noted, "which I passed with fair but not notable distinction at the end of the winter session of 1881. I was now a Bachelor of Medicine and a Master of Surgery, fairly launched upon my professional career."

For the moment, his Arctic journal was forgotten, but the experience had entered his bones. Seen at a remove of 130 years, Conan Doyle's diary aboard the *Hope* offers a remarkable window on the past. His is not only a coming-of-age story, but also a tale of Arctic exploration and a chronicle of a life at sea that no longer exists. To the end of his life, Conan Doyle would look back on the experience with awe. "You stand on the very brink of the unknown," he declared, "and every duck that you shoot bears pebbles in its gizzard which come from a land which the maps know not. It was a strange and fascinating chapter of my life."

Jon Lellenberg & Daniel Stashower

Facsimile of Arthur Conan Doyle's diary
of his voyage as ship's surgeon aboard
the Arctic whaler S.S. *Hope*,
February 28–August 11, 1880

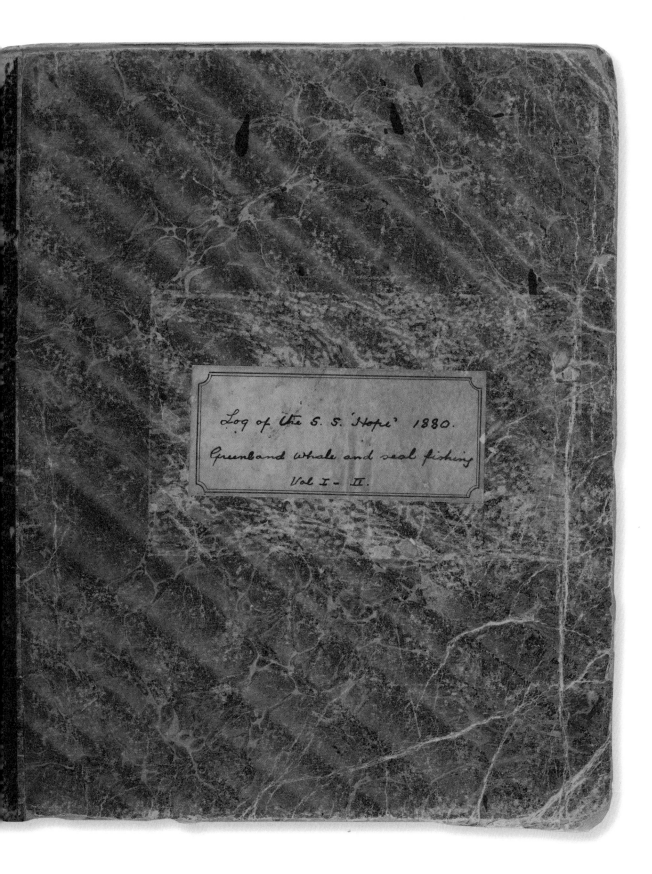

Log of the S. S. 'Hope' 1880.

Greenland whale and seal fishing

Vol I - II.

.....	5	12	19	26	
.....	6	13	20	27	
[Th]ur....	7	14	21	28	
[F]ri. ...	1	8	15	22	29
[S]at.....	2	9	16	23	30

FEBRUARY.

Sun......		7	14	21	28
Mon.....	1	8	15	22	
Tues. ...	2	9	16	23	
Wed.....	3	10	17	24	
Thur....	4	11	18	25	
Fri.......	5	12	19	26	
Sat......	6	13	20	27	

MARCH.

Sun......		7	14	21	28
Mon.....	1	8	15	22	29
Tues. ...	2	9	16	23	30
Wed.....	3	10	17	24	31
Thur....	4	11	18	25	
Fri	5	12	19	26	
Sat......	6	13	20	27	

Tues. ...		6	13	20	27
Wed.....		7	14	21	28
Thur....	1	8	15	22	29
Fri.......	2	9	16	23	30
Sat......	3	10	17	24	

MAY.

Sun......	2	9	16	23	30
Mon.....	3	10	17	24	31
Tues....	4	11	18	25	
Wed.....	5	12	19	26	
Thur....	6	13	20	27	
Fri.......	7	14	21	28	
Sat... 1	8	15	22	29	

JUNE.

Sun......		6	13	20	27
Mon.....		7	14	21	28
Tues. ...	1	8	15	22	29
Wed.....	2	9	16	23	30
Thur....	3	10	17	24	
Fri......	4	11	18	25	
Sat......	5	12	19	26	

Tues. ...		6	13	20	27
Wed.....		7	14	21	28
Thur....	1	8	15	22	29
Fri.......	2	9	16	23	30
Sat......	3	10	17	24	31

AUGUST.

Sun.	1	8	15	22	29
Mon.....	2	9	16	23	30
Tues.....	3	10	17	24	31
Wed.....	4	11	18	25	
Thur....	5	12	19	26	
Fri.......	6	13	20	27	
Sat......	7	14	21	28	

SEPTEMBER.

Sun......		5	12	19	26
Mon.....		6	13	20	27
Tues.....		7	14	21	28
Wed.....	1	8	15	22	29
Thur....	2	9	16	23	30
Fri......	3	10	17	24	
Sat......	4	11	18	25	

Tues.....	5	12	19	26	
Wed.....	6	13	20	27	
Thur....	7	14	21	28	
Fri... 1	8	15	22	29	
Sat... 2	9	16	23	30	

NOVEMBER.

Sun......		7	14	21	28
Mon.....	1	8	15	22	29
Tues. ...	2	9	16	23	30
Wed.....	3	10	17	24	
Thur.....	4	11	18	25	
Fri.......	5	12	19	26	
Sat.......	6	13	20	27	

DECEMBER.

Sun.		5	12	19	26
Mon.....		6	13	20	27
Tues. ...		7	14	21	28
Wed.....	1	8	15	22	29
Thur....	2	9	16	23	30
Fri.	3	10	17	24	31
Sat.......	4	11	18	25	

Tues.....	4	11	18	25	
Wed.....	5	12	19	26	
Thur....	6	13	20	27	
Fri.......	7	14	21	28	
Sat.... 1	8	15	22	29	

FEBRUARY.

Sun.		6	13	20	27
Mon.....		7	14	21	28
Tues. ...	1	8	15	22	29
Wed.....	2	9	16	23	
Thur....	3	10	17	24	
Fri.......	4	11	18	25	
Sat......	5	12	19	26	

MARCH.

Sun......		5	12	19	26
Mon.....		6	13	20	27
Tues. ...		7	14	21	28
Wed.....	1	8	15	22	29
Thur....	2	9	16	23	30
Fri.	3	10	17	24	31
Sat.	4	11	18	25	

TABLE

Showing the number of days from any given day of one month, to the same day in any other month. In leap-year add 1 if February 29 is in the calculation.

To the same day of	FROM ANY DAY OF											
	Jan.	Feb.	Mar.	Apr.	May	June	July	Aug.	Sept.	Oct.	Nov.	Dec.
January..........	365	334	306	275	245	214	184	153	122	92	61	31
February.........	31	365	337	306	276	245	215	184	153	123	92	62
March...........	59	28	365	334	304	273	243	212	181	151	120	90
April.............	90	59	31	365	335	304	274	243	212	182	151	121
May..............	120	89	61	30	365	334	304	273	242	212	181	151
June..............	151	120	92	61	31	365	335	304	273	243	212	182
July.............	181	150	122	91	61	30	365	334	303	273	242	212
August..........	212	181	153	122	92	61	31	365	334	304	273	243
September.......	243	212	184	153	123	92	62	31	365	335	304	274
October..........	273	242	214	183	153	122	92	61	30	365	334	304
November........	304	273	245	214	184	153	123	92	61	31	365	335
December........	334	303	275	244	214	183	153	122	91	61	30	365

DIARY FOR 56 YEARS, FROM 1844 TO 1899.

EXPLANATION.

Look for the [y]ear you want. Above it find its D[o]minical letter. Look on the lower p[ar]t of the Diary for the same letter in the line opposite the month wanted. Above that letter you find the days o[f] the week, and on the left the days of [th]e month. Leap years have two letters; the first serves till the end of February. Leap year is the year which divides evenly by 4; but the year 1900 will not be a Leap year.

GF 1844 1872	E 45 73	D 46 74	C 47 75	BA 48 76	G 49 77	F 50 78	E 51 79	DC 52 80	
B 1853 1881	A 54 82	G 55 83	FE 56 84	D 57 85	C 58 86	B 59 87	AG 60 88	F 61 89	
E 1862 1890	D 63 91	CB 64 92	A 65 93	G 66 94	F 67 95	ED 68 96	C 69 97	B 70 98	A 71 99

1	8	15	22	29	Su	Sa	Fr	Th	W	Tu	M
2	9	16	23	30	M	Su	Sa	Fr	Th	W	Tu
3	10	17	24	31	Tu	M	Su	Sa	Fr	Th	W
[illegible]	11	18	25		W	Tu	M	Su	Sa	Fr	Th
[illegible]	[illegible]	19	26		Th	W	Tu	M	Su	Sa	Fr
[illegible]	13	20	27		Fr	Th	W	Tu	M	Su	Sa
[illegible]	[illegible]	21	28		Sa	Fr	Th	[illegible]	Tu	M	Su

PUBLIC HOLIDAYS.

LONDON—*Exchequer, and India House.*—Good Friday and Christmas.

Law Offices.—Good Friday and three following days, Whit-Monday and Whit-Tuesday, Queen's Birth-day and Accession, Christmas, and three following days.

Excise, Stamp and Tax Offices.—Good Friday, Queen's Birth-day, June 28, Nov. 9, and Christmas.

Docks and Customs.—Good Friday, Queen's Birth-day, and Christmas.

IRELAND—*Customs.*—Good Friday, Queen's Birth-day, and Christmas.

Excise, & Stamp Offices.—Good Friday, Queen's Birth-day, June 28, Nov. 9, and Christmas.

SCOTLAND—*Customs*—Good Friday, Queen's Birth-day, Sacramental Fasts, & Christmas

Excise and Stamp Offices.—Jan. 1, Good Friday, Queen's Birth-day, (Fair Saturday, Glasgow), Sacramental Fasts, and Christmas.

BANK HOLIDAYS.

ENGLAND AND IRELAND.

Good Friday, Easter Monday, the Monday in Whitsun week, the first Monday in Aug., Christmas, & the 26th of Dec. if a week day.

SCOTLAND.

New Year's Day, Christmas Day, (if either of these fall on a Sunday the following Monday shall be a Bank Holiday,) Good Friday, the first Monday of May, the first Monday of August, and any day which may be appointed by Royal Proclamation. Upon Sacramental Fast-days and other local Holidays, the Bank Offices will be open only between the hours of 9 and 11 a.m.

POSTAGES.

INLAND LETTERS AND PACKAGES.

The rate of postage to be prepaid on inland letters and parcels of all sorts, closed or open, is as follows:—

Not exceeding 1 oz. in weight, ...1 *d.*
Above 1 oz. and not above 2 oz., ...$1\frac{1}{2}$*d.*
Do. 2 " do. do. 4 " ....2 *d.*
Do. 4 " do. do. 6 " ....$2\frac{1}{2}$*d.*
Do. 6 " do. do. 8 " ....3 *d.*
Do. 8 " do. do. 10 " ....$3\frac{1}{2}$*d.*
Do. 10 " do. do. 12 " ....4 *d.*

A letter above the weight of 12 oz. is liable to the charge of 1*d.* for every ounce, commencing with the first ounce. For instance, a letter between 12 and 13 oz. weight must be prepaid 1*s.* 1*d.* As a general rule, the postage if not paid in advance, is double the foregoing; and if the payment in advance be insufficient, double the deficiency is charged. An inland letter, for example, weighing more than an ounce, and not exceeding two ounces, and prepaid one penny only, is on delivery charged *double* the deficiency of one halfpenny, viz. one penny, and so on.—A letter directed to a person, and the person not found at the address, and the letter re-directed, single additional postage only is charged. The Post Office cannot undertake the safe transmission of valuable inclosures in unregistered letters. Letters when once posted cannot be given back upon any pretence whatever. Letters to warm climates should be gummed or wafered, not sealed with wax, as the wax is liable to get melted, to the injury of other letters.

A postmaster is not bound to re-direct letters for a person temporarily leaving his home and not having a private bag or box, unless the house be left uninhabited, or the letters would be delayed in their transmission by being sent to the house to be re-directed there. In all cases of re-direction, a written authority, duly signed by the person to whom the letters are addressed, must be sent to the postmaster.

REGISTERED LETTERS.

An inland letter or book-packet can be registered on payment of 4*d.* in addition to the postage of such letter or packet. The full amount in stamps for postage and registration, must be on the letter or book-packet, and it will require to be posted half an hour before the closing of the box for the mail by which it is to be despatched, otherwise a late fee of 4*d.* will be charged till the closing of the letter-box. A letter to be registered should be presented at the window and a receipt obtained for it, and must not be dropped into the letter-box. Any letter marked "Registered" which shall be posted without at the same time being registered shall, if observed, be afterwards registered by the post-office, and be forwarded to its destination, charged with double the ordinary registration rate of postage. All letters containing coin are treated as registered, even though they be posted without registration, and are charged on delivery with a double registration fee, in addition to the ordinary postage. Any person sending money or jewellery in a letter, if lost or mis-delivered, has no claim on the Post-office. The most secure mode of sealing is first to wafer or gum the letter and then to seal it with wax. No article of a dangerous kind can be sent by post. Scissors, knives, [illegible] may be sent if packed in

BOOK PACKETS,

Or plain, written, or printed paper, without a cover, or in a cover open at the ends or sides, may be sent—

Not exceeding 2 oz. in weight, ... $\frac{1}{2}$*d.*
,, 4 oz. ,, ...1 *d.*
,, 6 oz. ,, ...$1\frac{1}{2}$*d.*
,, 8 oz. ,, ...2 *d.*
,, 10 oz. ,, ...$2\frac{1}{2}$*d.*
,, 12 oz. ,, ...3 *d.*

And $\frac{1}{2}$d. for every additional two ounces.

The postage to be prepaid by postage stamps affixed. If the whole postage be not prepaid, the unpaid part is charged double on delivery. The packet may contain any number of separate books, prints, photographs (when not on glass nor in cases containing glass), maps, parchment or vellum, either printed or written, including printed or lithographed letters, plain or mixed, but not letters, sealed or open, nor anything sealed or closed against inspection. Marking or writing allowed when not of the nature of a letter. An entry, merely stating who sends the book, or to whom it is sent, is not regarded as a letter. The name and address of the sender is not only permitted but recommended, so that if the cover come off, or for any other reason the packet cannot be forwarded, it may be returned. In books and prints, all legitimate binding, mounting, or covering of the same, or of a portion thereof, will be allowed; as also markers and rollers, (whether of paper or otherwise), and whatever is necessary for the safe transmission of literary or artistic matter or usually appertains thereto. For the greater security of the contents, the packet may be tied at the ends with a string The postmaster is authorised to cut the string, but must refasten the packet afterwards. No packet can be received if it exceeds 5 pounds in weight, 18 inches in length, and 9 inches in width and 6 inches in depth, except petitions to Parliament.

INLAND PATTERN POST.

All parcels are forwarded as the sender desires, whether closed or open; the weight is limited to 12 oz., and the size to 18 inches in length by 9 inches in depth and 6 inches in width. The postage is the same as that for letters, and must be prepaid in postage stamps.

INLAND MONEY ORDERS.

For sums under 10*s.*, - - - -	1*d.*
" 10*s.* and under £1, - - -	2*d.*
" £1, " " 2, - - -	3*d.*
" 2, " " 3, - - -	4*d.*
" 3, " " 4, - - -	5*d.*
" 4, " " 5, - - -	6*d.*
" 5 " " 6, - - -	7*d.*
" 6, " " 7, - - -	8*d.*
" 7, " " 8, - - -	9*d.*
" 8, " " 9, - - -	10*d.*
" 9, " " 10, - - -	11*d.*
" 10, - - - - - - -	1*s.*

No order to contain a fractional part of a penny. No money order can be issued unless the applicant furnish the surname in full, and at least one initial of the Christian name of the person or firm who sends the order, and of the person or firm to whom the money is to be sent, unless the remitter or receiver is a peer or a bishop, then his ordinary title is suffi-

Log of the "Hope"   Vol II I.

5 12 19 26 | Tues

aur.
ri.
at....

Sun...
Mon...
Tues..
Wed...
Thur..
Fri...
Sat....

Sun....
Mon...
Tues..
Wed...
Thur..
Fri....
Sat....

Showing
the sar
is in th

T
the s
day

January.

February

March ..

April.....

May .....

June.....

July.....

August ..

September

October ..

November

December

DIARY FO

EX

Look for the
it find its D
on the lower
same letter
month wante
find the days
left the days
have two let
the end of Fe
year which
the year 1900

GF	E	D
1844	45	46
1872	73	74
B	A	G
853	54	55
881	82	83
	D	CB
862	63	64
1890	91	92

8	15	22
9	16	23
10	17	24
11	18	25
8	19	26
3	20	27
21	28

Saturday February 28.

Sailed at 2 o'clock amid a great crowd and greater cheering. The 'Windward' Captain Murray went out in front of us, their captain bellowing 'port' and 'Starboard' like a Bull of Bashan. We set about it in a quieter and more business like way. We are as clean as a Gentleman's yacht, all shining brass and snow white decks. Saw a young lady that I was introduced to but whose name I did not catch waving a handkerchief from the end of the pier. Took off my hat from the Hope's quarterdeck though I don't know her from Eve. Rather rough outside and the glass falling rapidly. Beat about the bay for several hours and had dinner with champagne in honour of Baxter and grandees on board. Pilot boat came and fetched them all off at last, together with an unfortunate Stow-a-way who tried to conceal himself in the Tween decks. Sailed for Shetland in a rough wind, glass going down like oysters. As long as I stick on deck I'll do.

Sunday March 1st
Got into Lerwick at 7.30 PM. Deuced lucky for us as

a gale is rising and if we hadn't made the
land we might have lost boats and bulworks.
We were uneasy about it, but we sighted the
Bussay light about 5.30. Captain very pleased.
We got in before the Windward, though they had
5 hours start.

Monday March 1st

Blowing a hurricane. Windward got in at 2AM
only Just in time. The whole harbour is one
sheet of foam. Feel very comfortable aboard.
Have a snug little cabin. Telegraph gone wrong
between this and Peterhead. Rokey holes.

Tuesday March 2nd

Glass down at 28.375. Captain has never seen it
so low. Blowing like Billy outside. Made out
the hosiery list. Tait on religion and atheism.
He is our Shetland agent, not half such a fool
he looks.

Wednesday March 3d.

Fine day- Glass still very low. Went on shore
with the Captain after breakfast. Enlisted our
Shetland hands. Fearful rush and row in Taits
small office. 'Jan Meyen' & 'Victor' came in. Murray of

the Windward seems a decent fellow. Captain and I were going from Tait' shop when a drunken Shetlander got hold of him. "Cap'n. I'm (hic!) goin' with you. Oh such a Voyage, Captain, such a voyage as never was landed! hic(!) three hundred and fifty tons, sir, I've brought the luck with me". Gray turned back in the back room, and seemed annoyed. I said "I'll turn him out of you like, Captain." He said 'Ah, I know fine you'd like a smack at him, Doctor; I would mysel but it wouldna do." We had locked the door of the back room when there was an apparition of a hand and arm through the smoked Glass window which formed the upper half of the door. Bang! Crash! Wood and Glass came rattling into the room, and we saw our 'indomitable Shetlander, with his hands cut and bleeding looking through the hole. "Wood and iron won't keep me from you, Captain Gray. Go I will." The Captain coolly smoked his pipe the whole time and never moved from the Stove. The man was carried off, kicking & thumping to the Gaol, I suppose, though the infirmary would have been a far better place. If ever I saw W & that was it. Had Murray of Windwar and Tait to dinner, talk of masonry, whaling &c. Dispute with Murray about efficacy of drugs. Plen of good wine going. Finished the evening with th Captain very pleasantly. By the way another Stowaway turned up today, a wretched looking animal, The Captain was frightening him after by telling him he'd have to go back, but he finally signed the Articles

Thursday March 4th

Gave out Tobacco in the morning. Slept forenoo Went ashore in the evening. Went first with second mate and Stewart to the Queens an had something short as he calls it. Then went to Mrs Browns and lost sight of them. Had a very hospitable reception there. Told me to make their home my home. Went down to the Commercial where an F & E was going on Heard some good songs and sang Jackis yarn Chat with Captain about Prince Jerome. &c.

Sunday March 6th

Captain and I were invited to Taits for dinner.
Both thought it a horrid bore. Went to the Queen's
and played billiards. Then toddled down to Taits.
met Murray of the Windward and Galloway, the
latter a small lawyer, insufferably conceited —
hate the fellow. Had a heavy weary dinner with
very inferior champagne. Old Tait expressed great
surprise at my saying I was RW's nephew — the old
cow, I found out afterwards that the Captain had
just been telling him about it. He has a dog
who has been taught to love the name of Napoleon,
if you talk of shooting. Napoleon he will make a
dart at you, and probably leave with some things
of yours in his mouth, muscles and clothes and
things. Murray talked about putting three men
under the ice, seeing ten men shot in a mob now
and several curious things. We got the boat
at nine o'clock and were both delighted to get
on board again, and stretch our legs quietly.
Wind rising. Saw what the Captain says is a
Roman Camp, but I think it is a Round Pictish
tower.

Saturday March 6th

Raining and blowing hard. Did nothing all day. Colin McLean and men went ashore in the evening and hailed for boat which we had to give though it was rather rough.
Began Boswell's life of Johnson.

Sunday March 7th

Nothing doing except that the mail steamer St Magnus came in with a letter from home and one from Letty, also a week's Scotsmen. Satisfactory news. We shifted our berth the other day in the harbour and now lie apart from the other ships with the Windward. Colin the mate was at the Queens last night among a lot of Dundeemen who spoke of those two D-d Peterheadsmen who went and moped by themselves. Colin got up and after proclaiming himself a "Hope" man ran a muck through the assembly knocking down a Dundee Doctor. He remarked to me this morning when I was giving him a pick me up "It's lucky I was sober, Doctor, or I might have got into a regular row." I wonder what Colin's idea of a regular row is. Lerwick is a dirty little town

with very hospitable simple inhabitants. Main Street
was designed by a man with a squint, builded
on the lines of a cork screw. Noticed today that
some of the ships in harbour flew Free mason
flags, Murray has the Royal arch up, Compasses
on a blue ground. Fishermen sell cod
here at 6/ a hundred weight, and
have caught as much as 25 Cwt in
a night. By the way the Engineer of the Windward
got his two fore fingers crushed in machinery
yesterday and I had to go over before breakfast
and dress them. Twenty sail of whalers in the
bay.

Monday march 8th.

Nothing like a quill pen for writing a Journal
with, but this is such a confoundedly bad one.
Went ashore today and after knocking about some
time went up to see a Football match between
Orkney and Shetland – Play rather poor. Met
Captains of Jan Mayen (Deuchars), Nova Zembla,
and Eric, also a London man, Brown, Doctor of
the Eric. Six of us went down to the Queens
after the match and started on bad whiskey

and went, to Coffee. Then Brown ordered a bottle of Champagne, and Murray and I followed suit. Cigars and grapes. I think we all had quite enough liquor. Brown was wrecked in the Ravenscraig last year. Says he is a very superior sort of shot. Captain and I got home about half past nine.

Tuesday March 9th

Went ashore with Captain before dinner. Jack Wilson was drunk and playing old Harry in the streets. Captain got hold of him and sent him on board the Hope in the pilot boat, but when he got half way he sprang over and swam ashore again. Cain and a boats crew captured him afterward. Had a very dull morning going from shop to shop. We will sail tomorrow if it is any way fair. Tait came on board afterwards and we had a pleasant talk. He is a sensible fellow tho' rather a bore. Looked over Scoresby. Captain told me some curious things about whaling. The great distance at which they can hear a steamer and how it frightens them out is about £50 a ton and bone £800 or so.

all bone goes to the continent. Sea Unicorns are very common, as are sharks, and dolphins but the curiosity of the place are the Animalcule which the whale eats.

Wednesday. March 10th

A North Wind prevented our getting off. The old Eclipse steamed in grandly about four o'clock being cheered by each ship as she passed. Went on board and saw Captain David, Alec and Crabbe. Went ashore in the evening and played Captain; also had the honour of beating Crabbe at billiards. He has a great local reputation. Left my meer--shaum and Gloves in the smoking room.

Thursday March 11th

A big day for Leith. The ships began to steer out from Lerwick sound after breakfast. It was a pretty thing on the beautifully clear and calm day to hear the men singing across the bay to the clank clank of the anchors. Every ship as it passed out got 3 cheers from all the others. Captain and I went ashore, and the boat's crew and I went in search of that beggar Jock Webster. We found him at last and five of us carried him, cursing horribly down the

31

main street of Lerwick to the boat, where I had to hold him to keep him from jumping overboard. We left about one o'clock and steamed through the islands till about seven when we came to an anchorage with the Jan Meyen, Eric, and Active in a little Voe. We raced the Jan Meyen up from Lerwick and beat her all the way, anchored within a stonethrow of the Eric. Talking to McLeod and Captain about getting to the pole in the evening. There is no doubt about it that every one has been on a wrong tack. The broad ocean is the way to find a way up to the pole, not by going up a drain which gradually grows narrower, and down which the ice naturally runs, as it does in Davis' Straits.

Friday March 12th

We'll have to stay here all day, I fear, for it is blowing half a gale tho' the glass is high. Nothing to do all day. The land is a succession of long low hills with peat cuttings and funny little thatched cottages here and there. Captain went over to the Eric in the evening. They seemed to be catching fish but we had no proper bait, so mate and I

went ashore with a boats crew to get some clams.
It was nearly dark so we couldn't gather again
but we went the round of the little cottages begging.
Such dismal hovels, the esquemeaux have better
houses. Each has a little square hole in the ceiling
to let out the smoke of a large peat fire in the
middle of the room. They were all civil enough
I met one rather pretty but shy girl even in this
barbarous spot. Got some razor fish as bait
and departed triumphant. Up to our thighs in
mud coming and going.   Revenue Cutter
boarded us this evening and Lieutenant was
only pacified by the present of a stick of baccy.
I'm afraid Colin will eat all our bait. Captain
rather annoyed about being kept in this hole.
Glass high
Saturday March 13th.
Wind high and raining hard. Active and Jan Meyen
are off already. We follow them soon. They are
pulling up the anchor now and singing "Goodbye,
Fare-thee-well, Goodbye Fare-thee well". A pretty
song it is too. Sea was not very rough outside –
Went through the islands, keeping Spell at the right

33

at the extreme north of Shetland we passed some curious rocks in the sea called Ramna Stacks.

Ramna Stacks.

Raining hard all day. We raced with the Eric and had rather the best of it. Not a bit seasick. Saw Burrafiord Holm the extreme north point of Great Britain, and then lost sight of land about four P.M. Ran with an oblique wind and three quarter steam all night. Dreamed of being beaten by a Gorilla, and of pulling in the Oxford boat. 167 miles

Sunday March 14th

Eric rather ahead of us and only occasionally in sight. Heavy Atlantic swell doing the Grand.

Heavy Atlantic Swell.

Northward Ho! all day under steam and sail. Northward Ho! ran about 150 miles. About getting to the Pole the gulf stream runs up past Spitzbergen so of course that is the way to go. It is one of the most extraordinary delusions in history how ship after ship has run up into a cul-de-sac. for Davis' Straits is nothing better. Read Boswell. Don't agree with McCauley at all about Boswell being a man of no intellect. If ever a man was afflicted with what he calls 'morbus Boswellianus' It is Lord McLauley himself in the case of Willy the Silent.

Monday March 15th

First under Steam and Sail, and then under Sail alone. Must have got about half way today. Kept in the cabin until evening. Read Boswell. Like that old boy Johnson for all his pomposity. A Thorough old fellow, I fancy. He was in Plymouth, it seems, for a couple of days, and there was considerable ill-feeling between the townsmen and the men about the docks. Johnson who had nothing in the world to do with it was often heard to exclaim "I hate a docker" I like that sort of thing. Sky looked like ice this evening. Surface temperature fallen from 44 to 38 in one day.

Tuesday March 16th

Still under canvas, wind continues fair. I've brought the luck with me. Two bottle nose whales were playing round the ship in the morning but I did not see them. It seems we are crossing a very favourite feeding ground of theirs. Expect to come on the ice tomorrow. We made 159 miles yesterday.

are hundreds of miles north of iceland, about sixty south east of Jan Meyen. Old hands on board say they never knew such a good passage, however we mustn't crow until we are out of the wood. Water temperature has fallen 2° since 12 o'cl. which looks like ice. White line on the sky. Everyone seems to think we will see ice before tomorrow. We can tell that we are under the lea of ice by the calm. Captain told me about some curious dreams of his, notably about the Germans and the black heifers

    Wednesday March 17<u>th</u>

Dies cretâ notanda. About five o'clock I heard the second mate tell the Captain, that we were among the ice. He got up but I was too lazy. Passed a Norwegian about 8 o'clock. When we rose at nine the keen fresh air told me it was freezing. I went on deck and there was the ice. It was not in a continuous sheet but the whole ocean was covered with little hillocks of it, rising and falling with the waves, pure white above and of a wonderful green below. None were more than 4 or 6 feet out of the water but they were of every shape. No Seals. Put up the crow's nest in the morning.

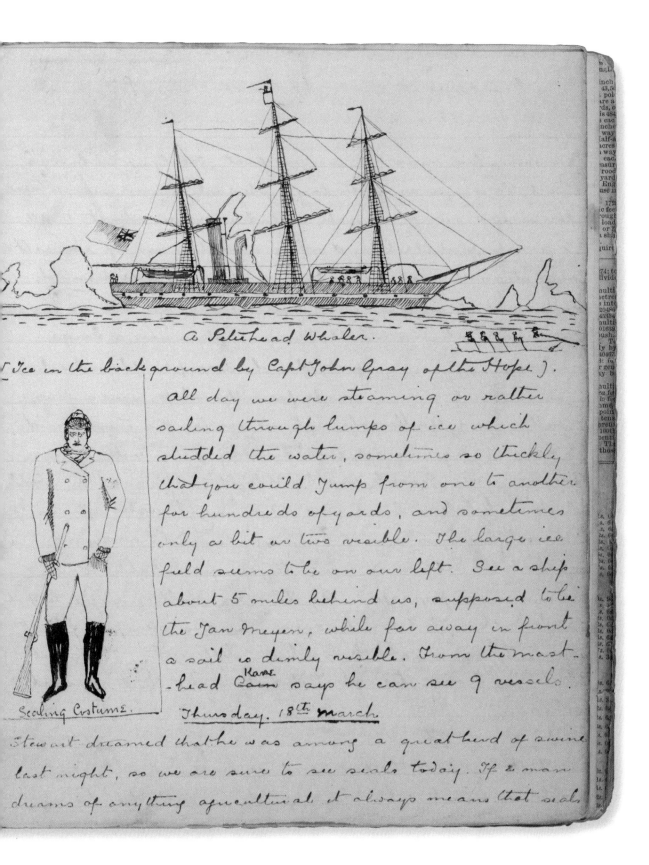

A Petehead Whaler.

[Ice in the background by Capt John Gray of the Hope].

Sealing Costume.

All day we were steaming or rather sailing through lumps of ice which studded the water, sometimes so thickly that you could Jump from one to another for hundreds of yards, and sometimes only a bit or two visible. The large ice field seems to be on our left. See a ship about 5 miles behind us, supposed to be the Jan Meyer, while far away in front a sail is dimly visible. From the mast-head Cain says he can see 9 vessels.

Thursday. 18th March.

Stewart dreamed that he was among a great herd of swine last night, so we are sure to see seals today. If a man dreams of anything agricultural it always means that seals are somewhere near. A curious fact. Ice lying in lumps much the same as yesterday. Stewarts dream seems true for we saw our first seal, a bladder nose, about 11 A.M. It was speckled black and white and lay on the ice as the ship steamed past, only about a dozen yards from it, looking at it quietly. Poor brute if they are all as tame it seems a shame to kill them. Captain saw a large speckled owl a couple of hundred

our first seal yards from the ship, saw a few roaches and Guillemots but we are too far from land to have many. We are considerably to the North of Jan Meyen now. Passed another bladder nose and a saddle-back seal later. Some were seen in the water afterwards. A most lovely morning but hazy towards evening. Spoke to the Eric and mutually congratulated each other on our passage. By the way Walker said to me at Lerwick "If I had known who you were, sir, last year, things might have been different.' I'm a lot better as I am, though I didn't make that remark to him.

Friday March 19th

A Thick haze with the lumps of ice looming out of it. Could see about a hundred yards in each direction.

38

passed two large bladdernoses, male & female on a bit of ice. We tried the whistling and certainly the male did stop and listen to it, the female wasn't so susceptible but shunted at once. The male was about 10 feet long I should think, the female 7 or 8. I wish the haze would clear up. Drizzling a little. . Haze continued all day so we lay to at night. Cane and Stewart were sparring in the evening. Talk on literature with the Captain, he thinks Dickens very small beer beside Thackeray. Buckland seems to be a lively sort of cove.

Saturday March 20th

Only a week from Shetland and here we are far into

are somewhere near. A curious fac[...]
much the same as yesterday. Sle[...]
for we saw our first seal, a blad[...]
It was speckled black and white a[...]
ship steamed past, only about a[...]
looking at it quietly. Poor brute[...]
it seems a shame to kill them.

speckled owl
our first seal. yards from
roaches and Guillemots but we[...]
to have many. We are considerably to the North of Jan
Meyen now. Passed another bladder nose and a saddle
-back seal later. Some were seen in the water afterwards.
A most lovely morning but hazy towards evening.
Spoke to the Eric and mutually congratulated each
other on our passage. By the way Walker said to me
at Lerwick "If I had known who you were, sir, last
year, things might have been different." I'm a lot
better as I am, though I didn't make that remark
to him.

Friday March 19th

A thick haze with the lumps of ice looming out of it.
Could see about a hundred yards in each direction.

The Hope [...]

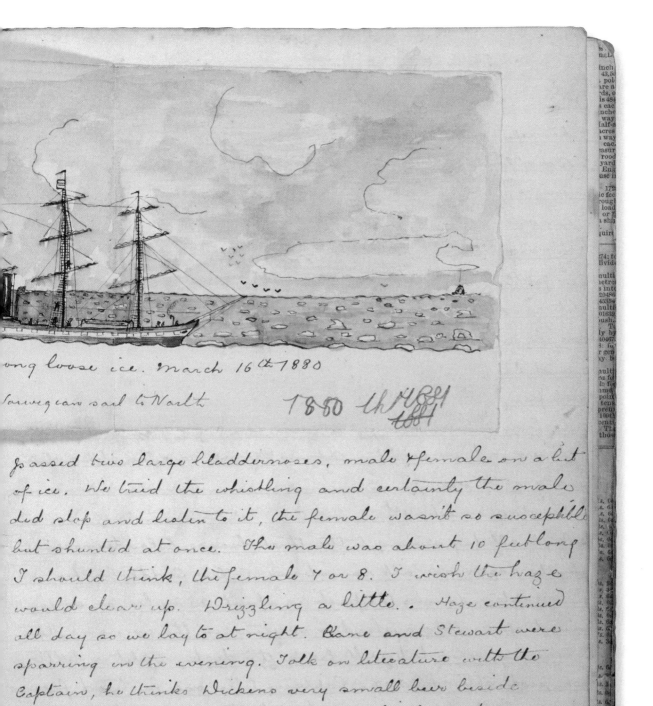

ong loose ice. March 16th 1880

Norwegian sail to North

1880 th 1881
1881

passed two large bladdernoses, male & female on a bit
of ice. We tried the whistling and certainly the male
did stop and listen to it, the female wasn't so susceptible
but shunted at once. The male was about 10 feet long
I should think, the female 7 or 8. I wish the haze
would clear up. Drizzling a little. . Haze continued
all day so we lay to at night. Bane and Stewart were
sparring in the evening. Talk on literature with the
Captain, he thinks Dickens very small beer beside
Thackeray. Buckland seems to be a lively sort of cove.

Saturday March 20th

Only a week from Shetland and here we are far into

the icefields. It has certainly been a splendid voyage. Beautiful day, wonderfully clear. Ice fields, snow white on very dark blue water as far as the eye can reach. We are ploughing through in grand style. Five sail in sight, one the Eric. Stewart insists on my accepting a pretty Esquimaux tobacco pouch; I suppose he means it as a quid pro quo for the Pipe I gave him. No seals seen as yet. Got near heavy ice in the evening and lay to. Several bladder noses playing about the ship. About a couple of hundred seals visible from the Crow's nest, so we seem to be coming near the pack. Eleven sail in sight. Adam Carner saw the slips of a bear in the ice

Sunday March 21st.

Lay to all day owing to the thick haze. Bladdernoses by the dozen are all round us. A few Saddle backs. The captain thinks the pack is about 20 miles or so in front of us. Johnny had a meeting in the evening, the singing sounded well from the deck. Split a bottle of port after dinner. Captain tells me he tried fixing a cone full of prussic acid onto the end of the harpoon. He fired it into a finner from his small steamer. The brute went away at such a rate that it very nearly set the bows on fire by the friction. The line broke and it got away, but seems to have died for no dog fish were seen on the coast for some days. Many finners are 120 feet long. By the way Carner taught me some esquimeaux. amalang (yes). pioue (very good) piow small (bad) kisi= =micky (ice - dog - is bear).

Monday March 22nd

Very foggy again, but we have drifted among a few saddle backs with their little fat yellow off spring. Got the quarter boats out, and the rifles. A long time to wait yet, though, till April 3d, Saturday week. Fog lasted all day so that we lay to. Boxed in evening. Finished Boswell Vol I.   Dreamed of G. P.

Tuesday March 23d

Clear morning, a good few seals in sight. Eclipse came in at last, and Captain boarded it before dinner. Steamed a few miles in the right direction. Blowing a gale all day. 11 degrees below freezing point. Very cold wind. Rigging covered with ice. Climbed up to the Crow's nest before tea, but the Captain called me down just as I got up to it, as

Male, female & young Saddleback.

43

he thought I might get frost bitten. Got a fine pouch from Cane. Carner tells me at New Orleans before the war a dock labourer could make £1 a day. Now they make a dollar only. Captain saw blockade runners leaving Liverpool during the war, long spider like steamers of great speed, and painted the colour of the ocean. Cargo mostly quinine, needed hardly any crew. Glass rising again.

Wednesday. March 24th

Another big day for Seth. We have seen the pack, and an enormous pack it is too. I have not seen it from the nest yet but it extends from one side of the horizon to the other, and so deep that we can see no end to it. The nearer we steam towards it, the bigger it grows. Colin says he never saw such a one in his life. It is certainly the largest collection of big animals in the world at present, at least I know no other beast that goes in herds of millions, covering a space about 15 miles long and 8 deep. We ought to have a good voyage now, my old luck. All the ships are lying round now and taking up their positions. Windward steamed past us today flying her Jack, and dipped it as a salute. 10 days yet to wait. Oysters.

44

Thursday March 25th

Hurrah for a quill pen! 19° below freezing point this evening.
Have been taking up our position, and mounting boats and
cleaning guns all day. Edge of pack can be seen from the
budge now. Good many isolated ones about the ship. I can
hear the young ones squeaking as I write. It is a noise
between the mew of a cat and the bleat of a lamb.
They look a sort of cross between a lamb & a gigantic
slug. Our only fear now is that some of these great
blundering Norwegians or Dundeesmen go and put
their foot into it. If we get less than 50 tons I'll be
disappointed, if we get less than 100 I'll be surprised.
Captain is going to teach one to take the latitude and
longitude. Saw a clever couplet today

"Till Science, like a poultice comes,
To heal the blows of Sound":

Holmes' I think. Sported my sea boots today.

Friday March 26th

Frost still continues, 17° today, 20° during the night.
This is Just what we want to fill up gaps in the
icefield and make it safe walking. Steamed very
little. The mate says the seals are lying in an
almost solid mass. He says there are more than
in '55, and in that year 50 vessels were among

45

them, and all got filled. We are 33 vessels now all told so the prospect is cheery. Bar Earthquakes we'll make a voyage of it. It is very trying work waiting, though this close time is an excellent provision. The poor brutes used to be killed before they had pupped. Eclipse got a bear today, and we saw the steps of one on the snow beside our ship. They are cowardly brutes unless in a corner. Captain killed one once with a boat hook. Engineer told me how one chased a crew for miles across the snow once, and how they had to throw down article after article to engage his attention, so that they got to the ship nearly naked and in a blue funk. There is no specimen of a right whale in any British museum, except a foetus. Saw the young seals suckling today. Hurt my hand boxing with the Stewart. Stuffed old Keith's tooth, and cured young Keith's Colly wobbles. It seems to be the family's day out.

Saturday. March 27th

This day week is our day. Got my knife and my

sharpener today, and asked Carner to see about my club. Beautiful day, still lying on the skirts of the pack, all seem satisfied except the Captain and he grumbles a bit, but I think he is only joking. Saw another bear's footsteps. The Eclipse has killed two and we have never seen one. They tell me bears go in flocks of 20 or 30 very often. Rifles given out tonight. Steamed a little. Hoggie Milne better tonight. No news. Wrote my "Modern Parable".

Bear's Step.

Sunday March 28th

Hoggie bad again so I gave him some Chlorodyne. Captain went on board the Eclipse and in a little the boat came off for me for dinner. Had a very pleasant feed with good wine afterwards. The conversation turned upon the War, politics, the North Pole, Darwinism, Frankenstein, Free trade, Whaling and local matters. Captain David seems to take a sinister view of our case. Says we'll be lucky if we get 20 tons; he may say it, but I don't think he thinks it. Saw his bear's skins. By the way he told us some strange stories which I will try to write as he told them.

When I was a young fellow; he said "I happened to be in London with a gold watch and a good deal of money. I was at the Lyceum one night and wanted to get back to my lodgings in Holborn but wandered about a long time unable to find my way. At last I saw a respectable looking man and asked him the way to Holborn, adding that I was a stranger. He said he was going that way himself, and that he was Captain Burton of the 17th Lancers. We walked on together and Captain Burton by turning the conversation on the danger of carrying money about in London, learned about my watch and gold, and warned me against it. We shortly afterwards turned into an open door and the Captain said "What shall we have here, I'll have some Cognac". I said "Coffee is strong enough for me". The waiter who brought in the things was the most repulsive looking ruffian I ever clapt eyes on, and I saw him stick his tongue in his cheek and leer at the Captain. It was then that I first suspected that I had got into a trap.

I threw half a sovereign on the counter and rose to go out, but the waiter put his back against the door and said "We don't allow our visitors to leave us like this." The Captain said "Come on, sir, and we'll make a night of it; hullo give us some Sherry out of bin No 3". The waiter called "Janet" and a girl appeared rather pretty and very pale. He said "Bin No 3". The girl said "Surely, Surely you don't need that bin tonight. He said "do what you are told." As she brought in the wine she whispered to me "Pretend to sleep". I drank a little of the wine, but spilled most of it. Then I sank down & closed my eyes. Soon the two villains came over and whispered together, and one passed the candle over my eyes and said "He is off". They whispered a little again, and one said "Dead men tell no tales" The other said "Then we had better get the bed ready" And they both left the room. I flung open the window and was off down the street like a shot, and ran about half a mile before I saw a

49

hobby, and then I found it impossible, with my imperfect knowledge of London to find the house again. I heard no more of it. Get out another bottle of Port, Doctor? The conclusion of the story was considered to be a very able effort. He told us another story about how he acted as a Spy in the Boer service, and murdered 3 Kaffirs in their sleep, and shot a German through the body.

He saw a Walrus eating a Narwhal once. He is a fine fellow, and Dr Walker seems a very decent chap too. He thinks more whales are found at night than in the day, so when he gets North into the Twilight land, he has his breakfast at 10 P.M. Dinner at 2 in the morning. And Supper at 7 A.M. Then he sleeps all day. He says whales leave a very characteristic odour behind them, and you often smell them before you see them.

March 29th Monday

Our time is coming now. Thursday will

a driving snow. Nothing particular going on. Had a pleasant evening in the mate's birth. Songs all round. Sang "Jack's Yarn" "The Mermaid" and "Steam Arm". Good fun. By the way Colin the mate payed me a high compliment today. He said "I'm going to have every man working hard when we start sealing. I've no fears of you, Surgeon, I'll back you to do a day's work with any man aboard. You suit one, and I liked the style of you the first time I saw ye. I hate your clean-handed gentlemen". This was a high com--plement from taciturn Colin.

March 30th Tuesday

Nothing much doing. Wendward came along--side and Murray came on board. He seemed to have small prospects, 10 tons was more than he expected, he said. Told us about Sir John Ross firing his gun through the window of a house because his mate was inside & he wanted him. Murray was one of the Franklin searchers. Ross said "Every step onwards, boys, is honour and glory to us. Death before dishonour", when they were starting sledging. Sparred with Colin & Stewart.

Wednesday. March 31st

Very little doing all day. A heavy swell has set in and we are uneasy about the result. If it continues until Saturday it will make our work both difficult and dangerous. The ice is not a solid sheet, but made up of thousands of pieces of all sizes floating close to each other. Now in a swell these pieces alternately separate and come together with irresistible force. If a poor fellow slips in between two pieces as is easily done, he runs a good chance of being cut in two, as actually happened to several Dundeesmen. Men played leap frog on a big piece. I started a story "A Journey to the Pole", which I intend to be good. We are going to write to Gladstone and Disraeli when the Dundeesmen go home.

Thursday April 1st

Swell continues and things look badly. We steamed a bit during the day. This is the first time for 3 years that I have not been examined today. Sent the Chief Engineer to the Captain with a cock and bull story about curtain rings. Johnny's dignity was very much hurt. By the way I was at the mast head yesterday, and also on the ice some time. Saluted the Harold Haarfager tonight 7.30. Swell still on.

Wednesday. March 31st

Very little doing all day. A heavy swell has set
in and we are uneasy about the result. If it
continues until Saturday it will make our work
both difficult and dangerous. The ice is not
a solid sheet, but made up of thousands of
pieces of all sizes floating close to each other.
Now in a swell these pieces alternately separate and come together with irresistible force.
If a poor fellow slips in between two pieces
as is easily done, he runs a good chance of
being cut in two, as actually happened to
several Dundeesmen. Men played leap frog
on a big piece. I started a story "A Journey
to the Pole", which I intend to be good. We are
going to write to Gladstone and Disraeli
when the Dundeesmen go home.

Thursday April 1st

Swell continues and things look badly. We
steamed a bit during the day. This is the
first time for 3 years that I have not
been examined today. Sent the Chief Engineer
to the Captain with a cock and bull story

ain rings. Johnny's dignity was very
. By the way I was at the mast
day, and also on the ice some
led the Harold Haarfager tonight
still on.

Ships taking up their positions among the seals – Birds eye view. march 26th 1880

Friday April Second

Swell still on and the pack growing more scattered.
I'm afraid our prospects will not be realized. How-
-ever every man must do his best, and then we
can do no more. Stayed up until 12 o'clock to
see the close time out.

Saturday April 3d

Up at 2.30 AM. Swell still on, so as to make
good work impossible. Lowered away our boats
in the sludge about 4.30. I stayed aboard at
the captain's command much against my will
and helped as well as I could by pulling the
skins up the side. The old seals who can swim
are shot with rifles, while the poor youngsters
who can't get away have their skulls smashed
in by clubs. It is bloody work dashing out the
poor little beggars brains while they look
up with their big dark eyes into your face.
We picked the boats up soon and started
packing, that's to say all hands getting
over the ship's side and jumping along
from floating piece to piece, killing all they
can see, while the ship steams after and

Seal Club

sucks up the skins. It takes a lot of knack to
know what ice will bear you, and what not.
I was ambitious to start but in getting over the
ships side I fell in between two pieces of ice and
was hauled out by a boathook. I changed my
clothes and started again, & succeeded in killing
a couple of seals and dragging their remains
after I had skinned them to the ship's side. We
got 7 5760 seals today. Poor work, I believe
but we hope for the best. After all whales are
the things that pay.

Sunday April 4th

Working all day. I fell into the Arctic Ocean three times
today, but luckily some one was always near to pull
me out. The danger in falling in is that with a heavy
swell on as there is now, you may be cut in two
pretty well by two pieces of ice coming together and
nipping you. I got several diags but was laid
up in the evening as all my clothes were in the
Engine room drying. By the way as an instance
of abstraction of mind after skinning a seal today
I walked away with the two hind flippers in my
hand, leaving my mittens on the ice. Some of

our hands work very well, while others, mostly Shetlanders with many honourable exceptions, shirk their work detestably. It shows what a man is made of, this work, as we are often killing far from the ship away from the Captains eye with a couple of miles drag, and a man can skulk if he will. Colin the mate is a great power in the land, energetic & hard working. I heard him tell a man today he would club him if he didn't work harder. I saw the beggars often walk past a fine fat seal to kill a poor little "Toby" or newly pupped one in order to have less weight to drag. The Capt ain sits at the mast head all day, looking out with his glass, for where they lie thickest. Took about 460 today.

Monday April 6th

Went out with Colin this morning for some regular hard work but began proceedings by falling into the sea again. I had just killed a seal on a large piece when I fell over the side. Nobody was near and the water was deadly cold. I had hold of the edge of the

ice to prevent my sinking, but it was too
smooth and slippery to climb up by, but
at last I got hold of the seals hind flippers
and managed to pull myself up by them.
The poor old 'flappy' certainly heaped coals of
fire upon my head. Got off again with the
Stewart and did some good work. Took
about 400 again.

Tuesday April 6th

Out on the pack in the morning with Colin
and actually did not fall in. The Captain
calls one 'the Great Northern Diver'. We
took a good number of young and old
and then steamed outside to see if we could
find anything for ourselves. Shot two large
bladdernoses, both were easy shots at
about 70 yards, but as I fired after all
the harpooners had missed I felt cocky.
They were huge brutes, I am keeping the bone
of one which was 11 feet long. They are also
called Sea Elephants. They have a vascular
bag on their snouts which they distend to
any extent when they are angry. Saw the

Jan Meyen and others, with all their boats out killing old seals. Took 270 young & 58 old

Wednesday April 7th

Poor work today, seals are scarce and we only book 133. Haggie Milne is very bad & I fear he will die. He has intus susception with faecal vomiting & constant pains. It is not hernia. Gave soap & castor oil injection today.

Thursday April 8th

Put our letters on board the 'Active' today. Had short notice and only wrote one letter though I would willingly have written more. Did a wretched days work, only about 30 seals. However most of the other ships have done worse than us, & that with crews of 80 men to our 56. Gale in the evening.

Friday April 9th

Gale continuing so that we have done no work at all. Heavy swell on. Got under the lee of the point. Wretched day. Did nothing but sleep & write up my log. They

are commencing to cut the blubber off the hides.
I'm afraid tomorrow will be as bad.

Waiting for the mother

Dragging Seals Skins

Clubbing a Young one.

a big load

my Accident

a Wasserous hit

Hauling a seal

April 10th Saturday

Poor Andrew Milne is almost beyond hope. At such an age and with such an illness recovery was almost hopeless. Blowing fitfully and with a heavy swell on. Nothing doing all day.

Began Carlyle's "Hero worship". A great and glorious book.

April 11th. Sunday

A dark day in the ship's cruise. Poor Andrew was very cheery and very much better in the morning, but he took some plum duff at dinner, and was taken worse. I went down at once, and he died within ten minutes in my arms literally. Poor old man. They were very kind to him forwards during his illness, and certainly I did my best for him. Made a list of his effects in the evening Rather a picturesque scene with the corpse and the lanterns and the wild faces around. We bury him tomorrow. Picked up seals all day on large pieces in the slush about 50 I think.

April 12th Monday

Buried poor old Andrew this morning. Union Jack was hoisted half mast high. He was tied up in canvas sack with a bag of old iron tied to his feet, and the Church of England burial service was read over him. Then the stretcher on which he was lying was tilted over and the old man went down feet foremost with hardly a splash. There was a bubble or two and a gurgle and that was the end of old Andrew. He knows the great secret now. I should think he would be flattened out of all semblance to humanity before he reached the bottom, or rather he would never reach the bottom but hang suspended half way down like Mahomets coffin, when the weight of the iron was neutralized. The Captain & I agree that on these occasions three cheers should be given as the coffin disappears, not in levity, but as a genial hearty fare - thee - well wherever you are. Did a fair day's work about 60 I should think. Made a bad miss in the evening. "Polynia" has 2050 seals, worse than us.

63

Tuesday April 13th

Boiled Beef day again (Tuesday - Tough-day - Tough day - Boiled Beef day). The worst dinner in the week except Friday. Lay to on account of the gale all day Had the gloves down in the Stoke hole in the evening and some fine boxing. No Seals.

IN. MEMORIAM. ANDR. MILNE
APRIL. 11th 1880

Wednesday April 14th

Knocking along among the ice under sail and canvas picking up seals. Made a good day's work, about 80 I should think bringing us up to 2460 about. Stood on the Folksel head all day and reported progress. Rather cold work had a shot or two tho? Some one told one that in the South Seas when a man died the first comer got his property, and that when a man fell over board you might see half a dozen standing by the hatchways to run down for the plunder whereever he was drowned.

Thursday April 15th

Beautifully fine day but we did a poor day's work, about 46 I think. Assisted in shooting 2 bladders. They took five balls each. A pretty little bird with a red tuft on its head, rather larger than a Sparrow came and fluttered about the boats. No one had ever seen one like it before. Rather a long beak, feet not webbed, white underneath, with a "pee-wheet - pee wheet": a sort of Snowflake. Georgey Grants got his trousers torn by a young Sea Elephant in the evening.

Friday. April 16th

Steamed hard to the North West all day to see if we could see anything of the seals. Failed in seeing many, and only picked up half a dozen. Jock Buchan shot a hawk in the evening which the Captain with his eagle eye discerned upon a hummock, and detected even at that great distance to be a hawk. About 18 inches high with beautifully speckled plumage.

my idea of a hawk. ~~(Angry Golden Age).~~ Had the Smallpox in its youth.

The Captain's idea of a hawk (N.B. Looking out for prey).

~~Happy and~~ The prey the Captain's hawk is looking out for.

Saturday April 17th

Nothing doing all day — only half a dozen seals again. We are steering South now with the "Iceberg" a Norwegian. If we could only make

Saturday's Night at Sea. April 17th /80.

it thirty tons I wd be satisfied. We have about 28 now I think. 26° of frost to day. Had singing in the evening in the mates birth. I Began a poem on tobacco which I think is not bad. I never can finish them. C'nest que la dernière pas qui conte.

Sunday. April 18th

A snowy drizzly kind of a day. Shot a seal in the morning off the bows; It was just sticking its head over the water. Saw two large sea birds, "Burgomasters" they are called. Went to a Methodist meeting in the evening conducted by Johnny McLeod the engineer, he read a sermon from an evangelical magazine and then we sang a hymn together. Argued afterwards with him.

Monday April 19th

Started Stuffing our hawk this morning, or rather skinning it for that is all I can do having no wires. I opened the stomach, then got out the legs to the knees and the humeri, and then inverted the whole body through the hole, cleaning out the brain, and removing everything except the skull. The result was satisfactory. We got a few bladders today, and are going North now to the old sealing. The Captain seems not to like the look of the ice at all.

A. Snap. Shot.

Tuesday
~~Sunday~~. April 20th

Nothing doing all day. Didn't take a single seal.
Sailed and steamed to the North East. 72 30 today.
Cleaned a couple of seals' flippers for tobacco
pouches, rubbed alum over our hawk's skin.

Wednesday. April 21st

Absolutely nothing to do except grumble, so we
did that. A most disagreable day with a
nasty cross sea and swell. No seals and
nothing but misery. Felt seedy all day. Was
knocked out of bed at 1 AM to see a man
forwards with palpitations of the heart. That
didn't improve my temper

Thursday April 22nd

A heavy swell still on. Took about 13
of which I shot two. Bad but better than
yesterday. Thick fog. Got a newly pupped
seal, it seems rather late in the season
for that. I have shot hitherto about 15
seals. I intend to count them after this.

Friday. April 23d

Did rather better today taking 36 seals. I
made a bag of 11, that is 26 altogether.

The shooting was uncommonly bad on the whole. Looks like a gale this evening. Captain saw another hawk. It is an extraordinary thing that we have not fallen in with a bear yet. Captain saw a meteoric stone fall into the water once within a hundred yards of the ship. The Magnetic Pole is in King Williams Land Lat 69°. There is another for South Pole, a thing that I never knew before.

Our Evening Exercise.

Saturday. April 24th

We have been steaming North West all day. Saw a fine flock of Eider ducks, the males are black and white, the females bronze with a green head. Picked up 17 more young seals. I think we are not very far from the old ones. Had a pleasant evening in the maters birth. No shooting today. Sparred in the morning. Have a tip to teach Jimmy.

was talking to Hutton, one of our best harpooners, about
zoological curiosities. He says that during a gale between
Quebec and Liverpool he saw two fish lying on the
surface of the water. They were about 60 feet long
and spotted all over, exactly like leopards. An
unknown Species. The Captain fell in with another
species in Lat 68° the hide of which was so thick
that no harpoon would pierce it. Here is my list
of Northern whales.

Right Whale.   proper Greenland whale. Yield
                    10-20 tons of oil. Bone sells at £1000
                    a ton. Value of one is £1500 - £2000
                    found in far North between the
                    ice fields
Finner Whale   Found in every sea in hundreds.
                    Are longer and stronger than the
                    Right whale. but very worthless.
                    Some 120 feet long. Razor Backed.
                    Spout two Jets, proper whale has
                    only one.
Bottle Nose Whales Found South of the ice, & round
                    Iceland. Only 30 feet long. Give
                    a ton of oil (£80). Skin Valuable.

White Whale (Beluga) found everywhere, including Westminster
Aquarium. Chiefly at mouths of American Rivers. Oil
valuable. 16 ft long.

Black Whale   a rare variety. Captain has only
seen one. Valuable. Americans
get them sometimes off North Cape.

Hutton's whale (Balæna variagatum)
Capt Gray's whale (Balæna ironsidum)

Sunday. April 25th

Got among a currant of young bladders in the
morning and took 22. I did good shooting before
dinner, hitting seven in eight shots from the bows.
Shot one after dinner and missed two which was
poor. We have 2502 now. Saw one old seal.
Boxed with Stewart and sang hymns with Johnny.
Drew a fine picture of Young Sealing.
Saw a good parody.

        "Oh the wild Rhymes he made,
            Small poets wondered
        To see in the 'Light Brigade'
            'Hundred' and 'Thundered'

Monday April 26th

Sailing N and NW all day trying for old seals.
They lie on the points of heavy ice stretching out
into the sea, but you never know exactly where
you can come across them, you must just

all h

coast along the heavy Greenland ice until you
find them. We are in 74° North today. Took one
young seal yesterday, and saw several. We
have nobler game in view. Boxed in evening.
Challenged Stewart to run a hundred yards.
I understand this sealing business thoroughly
now

White Whale (Beluga) found every where, including Westminster
  Aquarium. Chiefly at mouths of American Rivers. Oil
  valuable. 16 ft long.
<u>Black Whale</u>   a rare variety. Captain has only
                seen one. Valuable. Americans
                get them sometimes off North Cape.
Hutton's whale (Balæna variagatum)
Capt Gray's whale (Balæna Ironsidum)

  <u>Sunday. April 25th</u>
Got among a currant of young bladders in the
morning and took 22. I did good shooting before
dinner, hitting seven in eight shots from the bows.
Shot one after dinner and missed two which was
poor. We have 2502 now.   Saw one old seal.
Soued with Stewart and sang hymns with Johnny.
Drew a fine picture of young Sealing.
Saw a good parody.
        "Oh the wild Rhymes he made,
            Small poets wondered
        To see in the 'Light Brigade'
            'Hundred' and 'Thundered'
  <u>Monday April 26th</u>
Sealing N and NW all day trying for old seals.
They lie on the points of heavy ice stretching out
into the sea, but you never know exactly where
you can come across them, you must just

All hands over the bows — Young Sealing. 1880.

...eavy Greenland ice until you
...re in 74° North today. Took one
...terday, and saw several. We
...re in view. Boxed in evening.
...rt to run a hundred yards,
...is sealing business thoroughly
now

Point 4

Point 3. Where they are next

Point 2. where you will seek them
after they have been driven off
Point 1.

Heavy Greenland ice

[Several hundred miles
across]

Point 1. where you will find old ones

ARCTIC. OCEAN.

Jan Mayen

Point where Seals pup.
Young Sealing.
Old ones then go North

Plan of Greenland Seal fishing

Tuesday April 27th

Steaming N and NW all day. We have been
among young bay ice and are trying to make
the heavy where we may expect seals. Looks as
if we were not far off towards evening. Did
nothing all day. The skin of my hawk is just
ruined. Drew Milne's funeral again at night
for his brother who is aboard.

Wednesday April 28th

Made the heavy Greenland ice early in the morning
and when I came on deck after breakfast it was

stretching along the whole horizon. Heavier ice than any I have seen yet. The effects of the Arctic refraction are very curious. Here are two views at the distance of a mile and close up

Heavy ice close up.

Heavy ice at a mile

Saw marks of large herd of seals on the ice. a few in water steaming Northward. Was close to the Victor of Wunder in the evening. Beggar has no right to be there. Have the best prospects for tomorrow.

Thursday. April 29th

Our prospects have not been realized, for although we saw a few schools of seals in the water, we have not reached their head quarters though we have been steaming North all day. Capt Wavidson of the Victor boarded us this morning — a poor specimen of a man, hairy also. Was our mate once, and had the reputation of being a sulky beggar. The effects of refraction were extraordinary tonight, many pieces of ice appearing high up

in the air with the sky above and below them.
Victor steamed after us all day. We are not
far from the seals I'll bet. Saw many Snow
-birds about which is a good sign. In
Latitude 75° 11.

Friday April 30th

Morning broke very inauspiciously with a
Southerly wind and a hazy sky. We are beginning
to feel a bit downhearted as our sealing should
be begun by this time. Steamed North East after
the haze rose, water was like a lake with a
great deal of bay ice and numerous Looms
and Petrels. Just before tea we saw a point
of heavy ice ahead, and hope to find the seals
at the other side of it. I am rather doubtful
as we have seen none in the water as yet.
the night is very nearly as light as the day
now, I can read Chambers Journal at mid
-night easily. Served out Grog this evening
as tomorrow is the first of may. The ice
looks well for the whaling. 10.PM As I thought
there are no signs of seals upon the ice, so we
have come to the conclusion that we are probably

to the North of them. "Mine too mammy". They generally shave newcomers on the first of May and a boatsteerer told me this evening that I was to be a victim, but I told him they would have to call two watches to do it. ~~xxxxxxxxx~~ I ~~xxxxxxxxxxxxxxxxxxxxxxxxxxxxxxxxxxxx~~ ~~xxxxxxxxxxxxxxxxxxxxxxxxxxxxx~~ feel ~~xxxxxxxxxxxxxxxxxxxxxxxxxxxx~~ will wait till midnight ~~xxxxxxxxxxxxxxxxxxxxxxxxxx~~ Joby ~~xxxxxxxxxxxxxxxxxx~~.

Saturday May 1st

In the morning there was a heavy swell on and our prospects were of the darkest. Before dinner however ~~xxx~~ there was a change for we saw a young bladder on the ice, and shortly afterwards a considerable school of seals in the water going before the wind. The Captain and all of us were rather gloomy at dinner time, but the moment he mounted the Crow's Nest after dinner down came the welcome shout "Call all hands." I was in the Second mate's boat and we lowered away about 4.30. P.M. It was a fine sight to see the seven long whale boats springing through the blue water as we made for the ice. The seals seemed a good deal whiter than when I saw them last. Mate fired and missed, and not a shot could I get as he took the boatsteerer with his club and left me to find my way, and test the ice with the butt end of my rifle as best I could. Which was rather scurvy conduct. It was dangerous work on the ice as I could see no one, and twice only just saved myself from falling in when I should in all probability have been drowned. It is more dangerous work than the young sealing, for the sea undermines these lumps of heavy ice, so that when you think you are perfectly safe on a large piece the whole thing may crumble thro'. Never got a shot the whole time. Mate slew one. Most miserable work, the worst boat of the lot. If he could shoot as well as I, or I could walk on ice as well as him we would have had a different tale to tell. However it is jolly to get to work again.

seals were soon frightened off the piece. Went
off to the Westward.   Here is our day's fishing
Buchan's Boat.   14
Rennie's Boat 13
Carnie's Boat   10
Colin's Boat   10   Total 69.
Matheson's Boat 10.
Hutton's Boat 9.
McKenzie's Boat 2.
Cane's Boat 1

   Sunday May 2nd
Showers, heavy ice, snow and wind all conspiring
to ruin us. We steamed North in the teeth of a
gale all day persevering manfully. In the
evening the Captain came down from the mast-
-head almost in despair, and pointed out a
'blink' in the sky showing heavy ice ahead.
If the seals are not there', he said ' I must
turn South again? We steamed along and
then to my delight after tea 'All Hands'
were suddenly called. A considerable body
of seals were in sight but as others were
seen coming on to the ice, it was thought

advisable to leave them tonight & attack
them tomorrow. The great thing is to try
and get a turn at them before the 'Victor'
sees what we are doing. This Dundee
ship is the only one in sight. The Captain
of her is a sumph. Turned in early for
an early rise.

Monday May 3d

Boats lowered away about 6 A.M. The
moment they were down the Victor, who
is about five miles off, turned & steamed
furiously towards us. I went with
Peter McKenzie, the last of the harpooners.
We call our boat The 'mob'. It is manned
by all the rag tag and bobtail of the ship.
but I think has as good a crew as any.
There is Peter harpooner, Jack Coull Steerer,
The Doctor, Steward, Second engineer, and
Kieth the oldest man in the ship. We were
the last to leave the ship as the boats were
dropped one here, one there. The ice was
very heavy and good. at first we had
a bad berth and only shot 2 seals, but

we poked and pushed our way by sheer
hard work through the ice and got into a
fine bay lined with seals. Peter and I
sprang out with our guns and wriggled
our away along the ice, while the crew
crept after us to skin what we shot. I
saw Peter shoot two, and then I floored
one. Then I got behind a hummock and
shot nine, five all in a line on the edge
of one piece. I was just thinking we
would make a good bag and had shot
another, while I could hear the ring of
Peter's rifle a hundred yards off, when
in came the Victor's boats, pell mell all
in a heap right at the back of us. The
men sprang out, rushing across the ice
firing without aiming, jumping up on
the top of hummocks, shouting, and
making the most fearful mess of it.
They scared the seals and spoilt our
work and their own too. I don't suppose
they got fifty seals all together. Our boat
had 27 and the united total of our

morning's work was 234, Hutton heading
the poll with 68, and Cain having only 8.
Our Captain lowered the ensign 3 times to the Victor
as an ironical "Thanks for your politeness". The
moment the boats were aboard we set off at a
great rate as the Captain saw a fresh and a
larger batch of seals. We had a mouthful of
dinner it being 2. P.m. When we came on deck
after it we found the Victor had already landed
her boats, so choosing another spot off we
went. There was an enormous body of seals
but very shy, so that we had to make long
shots. We got 28 this time and the total
came to 287 or so. Altogether today we got
540 ‒‒‒‒, a splendid days work, about 11
tons of oil. Felt tired as I had been pulling
and crawling on my face all day. Captain
sees another patch of seals for tomorrow.

<u>Tuesday May 4<u>th</u></u>
at it at 6 AM again. Boats lowered away
and dropped here and there as usual.
Peter and I got behind a hummock and
shot 7 each, when the Captain saw he

had not got the thick of them, so he hoisted the Jack as a signal to the boats to return, he took the first five that came up, including our noble selves in tow, and away he steamed at full speed right past the Victor's boats and dropped us in among a fine patch. It was an energetic and sensible action. I suppose he towed us about 15 miles or so. We made good use of our chances then and shot away hard. The "mob" distinguished itself killing 41, Buchan was best with 75, then Colin with 51, Carner 42, We 41 and the rest very poor. We took 275 and did not lower away again. A norwegian ship the Diana came in on our flank but did not get very many. One of them boats came alongside ours and we asked them if they had seen the Eclipse to the North, they said they had, but I doubt if they understood us. Victor had its men out all night last night - a very short sighted policy.

Wednesday May 5th

Steamed to the N.E. Open water round us.

Steward.   Ego.

Five Bulls at a hundred yards
may 3d 1880.

Hardly expected any seals – All hands were
called however Just before dinner – The Diana
got the better of us rather, having all her boats
in the heart of the pack before we lowered away
The seals were lying very thick but not over
any great extent of ice. Our fellows muddled
it completely, each being anxious to get the
best position and beat the others. The seals
were finally scared off after we had taken
71. Captain seems displeased, and quite
right too.

End of Volume I. Log of Hope.

	may 1.	3	4.	5.	14	15
Colin	10	35	51	8	7	
Cane.	1	36	20	2	13	
Conner	10	61	42	11	2	
Hutton	9	112	11	6	12	14
Buchan	14	87	75	18	20	
Rennie	13	68.	26	7	2	
mulliison	10	47	10	11	2	
mckenzie	2	55	41	8.	5	
	69	540	275	71	63	32

Scrimmage. 56.

```
 69
540
275
 71
119
 32
------
1106
2502
3608
```

Game Bag of the "Hope". Voyage. 1880.

				my Ba
April. 3.	760. Young Seals.	50 Old Seals.	1 Old Sea	
4	450 Young Seals.	10 Old Seals.		
5	400. y. S.			
6	270 y. S.	6 Bladders. 57 Old Seals. 2 Bla		
7	133 y. S.			
8	30 y. S.			
9	50 y. S.			
10	72. y. S.			2 Seal
11				
12				
13				
14	80 y. S.			2 Seals
15	46 y. S.	2 Bladders.	2 Seals. 1 Blad	
16	6. y. S.	a Hawk.		
17	10. y. S.			2 Seal.
18	10 y. S.			1 Seal.
19	6. y. S.			
20				
21				
22	13 y. S.			2 Seals
23	36 y. S.			11 Seals

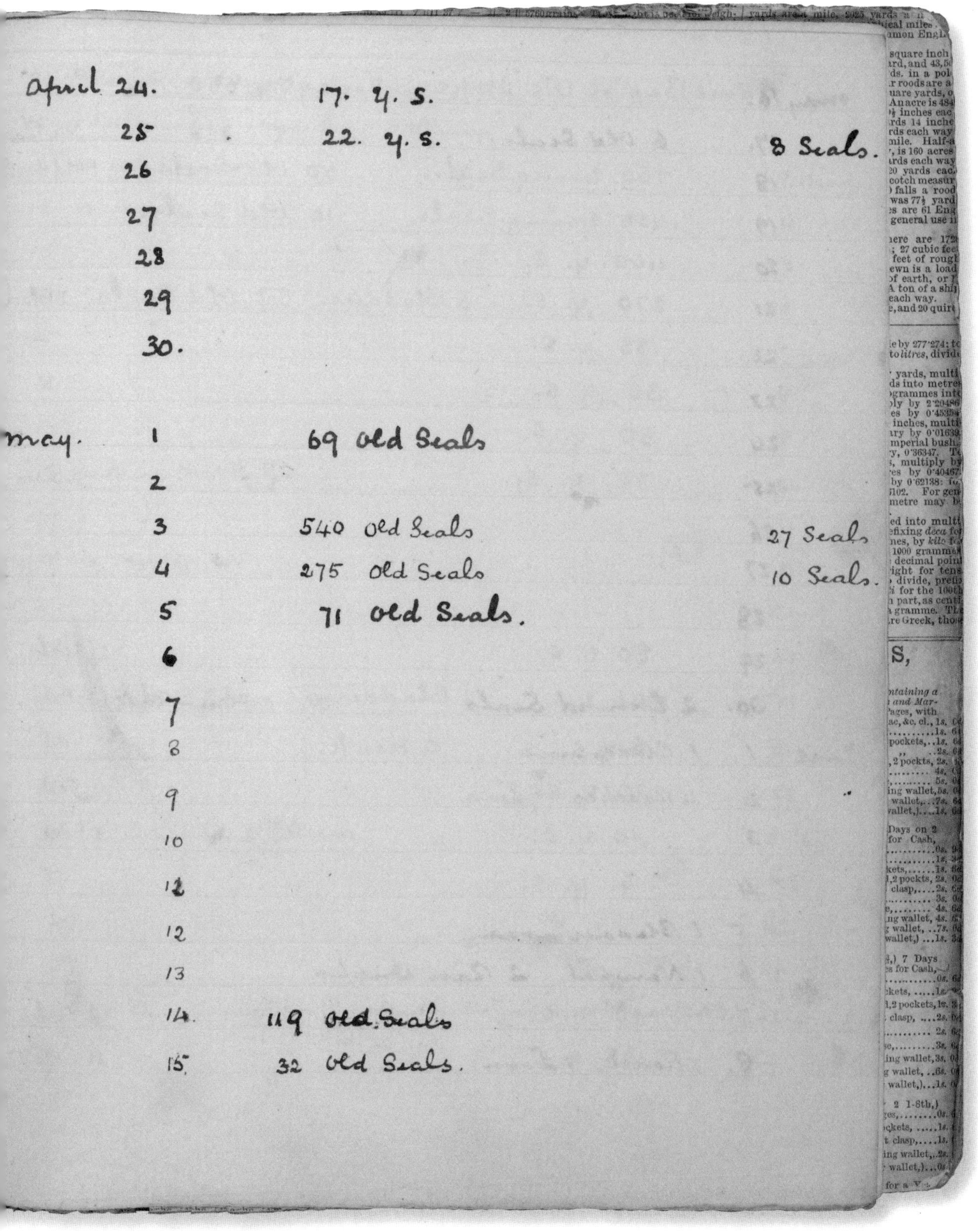

April 24.   17. 4. 5.

25   22. 4. 5.   8 Seals.

26

27

28

29

30.

May.   1   69 Old Seals

2

3   540 Old Seals   27 Seals

4   275   Old Seals   10 Seals.

5   71   Old Seals.

6

7

8

9

10

11

12

13

14.   119 Old Seals

15.   32 Old Seals.

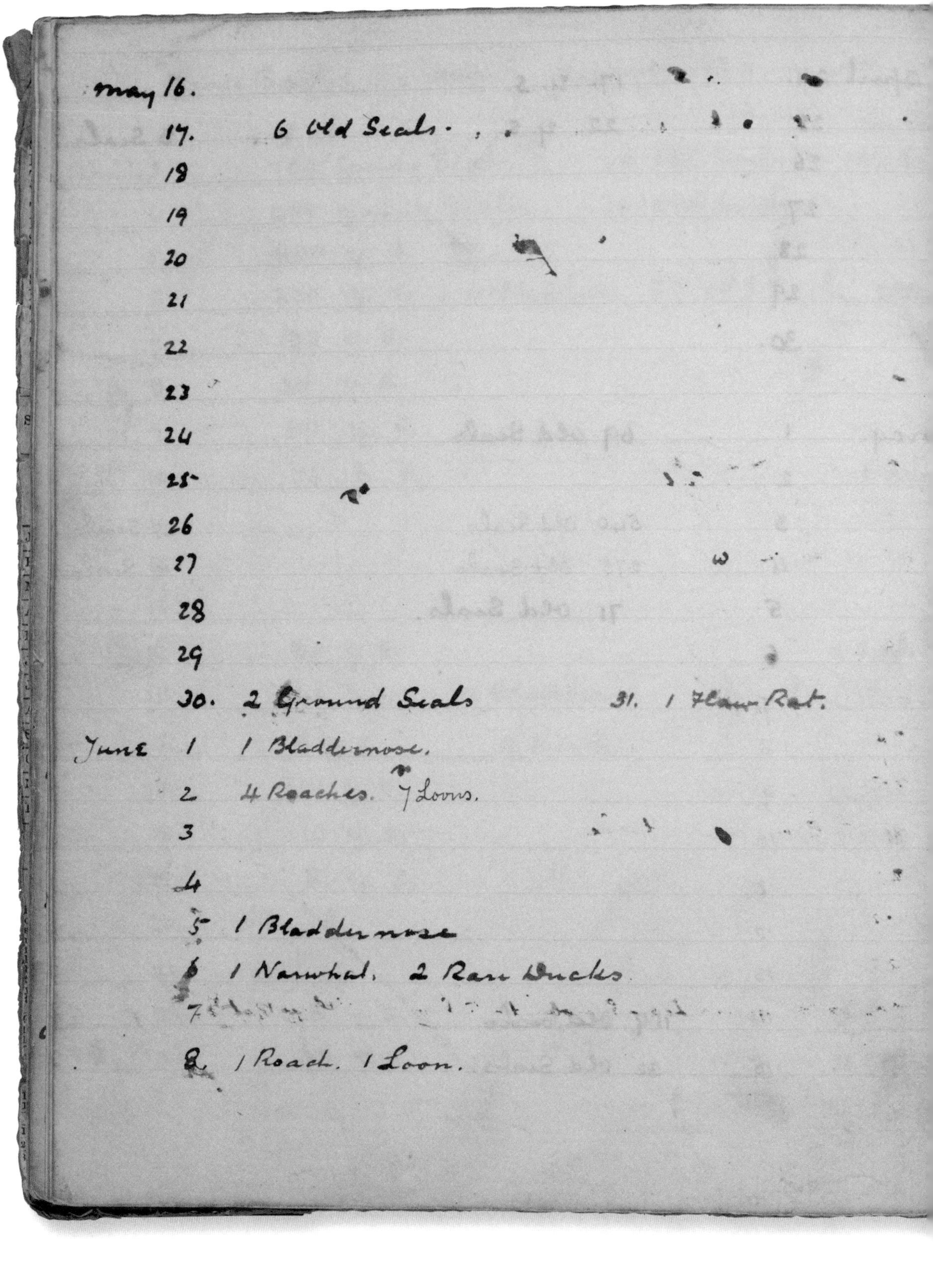

May 16.

17.   6 Old Seals.

18

19

20

21

22

23

24

25

26

27

28

29

30.   2 Ground Seals   31. 1 Flaw Rat.

June 1   1 Bladdernose.

2   4 Roaches. 7 Loons.

3

4

5   1 Bladdernose

6   1 Narwhal. 2 Raw Ducks.

7

8   1 Roach. 1 Loon.

9. 1 Roach. 6 Snowbirds

10 1 Kittewake. 1 Maulie. 3 Loons

11. A Whiting

12 A Bear

13

14

15

16

17

18 A Bear & 2 Cubs.

19

20 A Bear

21

22

23

24

25

26 A Greenland Whale

27

28

29

30. A Burgomaster - Snowbird - 5 Loons. 1 Haw Rat.

Log of the Hope Vol II.

May 6th Thursday.

The Captain has come to the conclusion that we had better go South a bit for seals, and let the Wiana and Victor go on to the North. We are in 77. 20 today, and expected to see the West coast of Spitzbergen. All hands are making off. there is a heavy swell and no signs of seals as yet.

Pull
Old Sealing

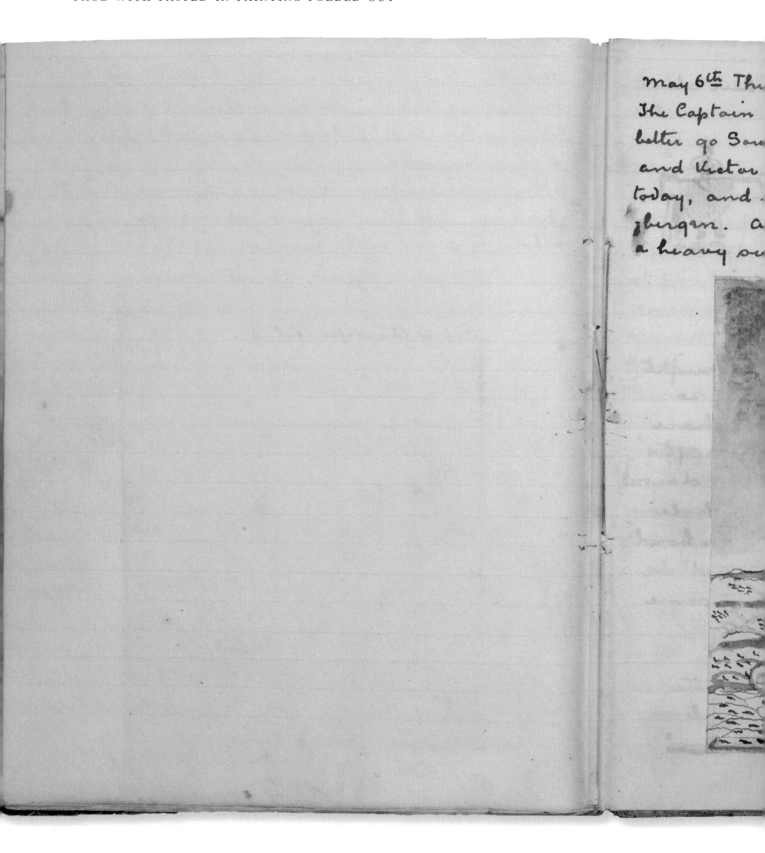

May 6th Thu
The Captain
better go Sou
and Victor
today, and
bergen. a
a heavy su

...day.
...come to the conclusion that we had ...a bit for seals, and let the Diana ...on to the North. We are in 77.20 ...ected to see the West coast of Spitz- ...hands are making off. there is ...b and no signs of seals as yet.

Pull, boys, pull. !
Old Sealing. Drawn in a heavy swell.

Steamed SW all day but saw no seals. We have about 50 tons now.

Process of making off in Sealing
(i.e. Separating the blubber from the skin).

Friday May 7th

The Diana seeing us lying to yesterday night thought we saw something splendid so she came down at a fearful rate to share the booty under steam and canvas. After burning 30/ worth of coals it began to dawn upon the Norwegian mind that the whole thing was a 'do' and a sell, so with a howl of disgust it flitted off again, to Iceland we believe. Under sail all day to N.E. Saw some schools in the water.

Saturday May 8th

Steaming N.W. Victor in sight going in the same direction. He follows us as a Jackal follows a lion. Ice all marked with seals. a beautiful day. Gave out tobacco and sugar in the evening. Was amused by a sailor's auction. Manson Turville a Shetlander was auctioneer & was particularly eloquent about a very delapidated and seedy old coat of his which he wanted to palm off. "Going at five plugs of tobacco, at five plugs! Nobody bid any more? a coat warranted to keep out anything under 190 degrees of frost – no advance on five plugs? Gentlemen! Gentlemen! Five plugs and a half. Thank you, sir! Going at five and a half! The figure of a beaver will be found on one side of the lining and a rattlesnake on the other. not sold but given away! Going. Going now or never. Gentlemen, now's your chance for a bargain. Gone." I bought a pair of sealskin trousers from Henry Polson.

Not Sold but Given away.

99

Sunday May 9th

Why are seals the most holy of animals? Because it is mentioned in the Apocalypse that at the last day an angel shall open six of them in heaven. None of them to be seen today. The thing is growing monotonous as Mark Twain said when the cow fell down the chimney for the third time while he was composing poetry below. A cloudy day. Have been reading Scoresby's book on whaling. Some of the anecdotes are too big to be swallowed at a gulp, they need chewing. He saw a whale caught in the bight of a rope that held another whale fast. Saw a man go a quarter of a mile on a live whale's back. However on the whole it is an eminently readable book, and very accurate as far as I can judge! Nothing all day. Was down in the Harpooner's berth in the evening, conversation ran on Zoology, murders, executions and Ironclads. Steaming to the Northwest.

Monday May 10

We are down in 73.20 now or only just to the N of the place where we were young sealing. We are going North again I am glad to say. No Seals.

served out tot coffee and tea in the morning. Glass had a tremendous fall after tea, and it came on thick rain and wind. I hope we are going to have a bit of a hurricane. Anything to wake us up. A codfish has been brought up through the pumps in a case of a big leak.

Tuesday may 11

A heavy gale during the night, and nearly all day. We hardly feel more than the force of the wind however, as the ice forms fine natural harbours. Running North all day. It is too bad this - after we began our old sealing so well too. However this is a trade of ups and downs, and we must wait for the swing of the pendulum. Old Peter got a nasty cut over the eye tonight from a rope, and seemed to think he was blinded, but I set him right again. Misery & desolation.

Wednesday may 12.

A most beautiful day. Blue sky which is rare up here as the sky is usually rather peasoupish. A good many seals in the water but none on the ice. As clear at midnight as during the day. A 'Finner' whale was seen spouting near the

ship after dinner, but I was asleep myself at
the time. I would have liked to have seen it.
Balvena Physalis is its scientific name, and it
is the swiftest, strongest, biggest and most
worthless of the whale tribe, so hunting them
is rather a losing game. However there is a
regular Finner fishery. They are worth about
£120 each, and our whale about £1500, so
we are on the right side of the bargain. Played
Catch the ten in the mate's berth for love. The last time I had
a card in my hand was at Greenhill Place. Saw a 'Flaw rat'
today swimming round the ship. It is the smallest variety
of seals. Captain's idea for the cure of baldness. Pick
hairs out of another man's head by the roots. Then bore
little holes in your head and plant them. He dreamed it.

Thursday may 13th

I hear from the engine room that Mr McLeod, our
chief engineer, has done me the honour to read my
private log every morning, and make satirical
comments upon it at table, and among his own
foremen. Now I would as soon that he read my
private letters as my Journal, in fact a good deal
sooner, and it is just one of those things which I

won't stand from any man. If any man meddles with my private business I know how to deal with him. I am only astonished that a man professing religious principles, should act with such a want, I won't say of gentlemanly honour, but of common honesty. If he does it after this warning he shall answer for it to me. A sensible man might be trusted, but a man who will talk about my prejudices against boiled beef &c in the engine room must be suppressed. I hope this may meet his eye in the morning.

Saw a 'Finner' whale today. I had no idea of the size and sleekness of the brutes before. His blast looked like a puff of white smoke. He was a good quarter of a mile from the ship but when he dived I could see every fin. A most enormous creature. Prospects look brighter this morning as seals have been seen on the ice, and a good many in the water. Cane came running down about 11. PM to say that he saw a good strip on the ice.

Friday may 14th

Boats lowered away about 9 A.M. Seals would not lie at all though. They have come up all the way from the

Labrador coast, and are nothing but skin and hair after their month's travelling. Our boat was one of the highest with 5. We took 63 in the boats & then came ~~aboard~~. The harpooners were sent over the bows to attack the remainder of the pack and killed 56 more, making a total of 119. It is pleasant to get started again. Cane was frightened by an enormous Walrus with a head like a barrel coming up beside him while he was flinching a seal on the ice. He fired 4 shots into it, but it only seemed amused, and swam away.

Poking the 'mot' boat through heavy ice. May 14th 1880.

Saturday may 15th

Sent away two boats in the morning, Colin's and Hutton's, to a small patch of seals. Took 32 of them. Steamed and Sailed North afterwards but saw no more blubber. The time is rapidly approaching now when we must coil our whale lines and go North

Unless we fall among seals in the next two days we must give it up. We think the eclipse and the Windward are North already. We have seen nothing of them since more than a month ago, when they ran down to Iceland for the bottlenosing. Reading Scoresby's Journal of his discoveries between Lat 69° & 76° on the coast of East Greenland. The lost Danish Settlements on that coast are a very curious problem. He found no trace of them.

Sunday May 16th

No seals. Lat 76.33 at noon. Banging away to the North as hard as we can go. Port Wine. Old Cooper tumbled down the hatch and broke his arm nearly.

Monday May 17th

A beautiful day. Steamed to the North all day. Lat 77½ Longt 5 East. About 100 miles West of Spitzbergen. Got 6 from a small patch of seals after dinner. They are getting very thin. The ones we have captured lately we consider to be, not Greenland seals, but seals from the Labrador coast which after their months travelling could hardly be expected to be in prime condition. Rigged up

the harpoon guns this evening. Fearfully cumbrous
things working on a swivel with a pull of 28 lbs.
Has to be let off by pulling
at a string. Carries a harpoon
about 30 yards with
some accuracy. Bore about 1½ inches.

Tuesday may 19th

Cleaned out the boats and
made all straight for whaling.
During dinner a sail was seen to
the N.W, which turned out on closer
Capt. J.G.   inspection to be the Windward, which we
thought was South with the Eclipse at the Bottle-
-Nose fishing. We hauled our yards aback and
waited for him. Murray came aboard with a
very dismal tale to tell. After the young sealing
he had been too ambitious to content himself with
the modest work that we had stuck to, picking
up half a ton a day or so, but he had run
right away North at once to Spitzbergen after
whales, not taking a single old Seal. The
result of his ambition is that we have about
52
½2 tons now, and he about 28. He has been

106

here three weeks and never seen a fish. He
gives a most discouraging account of the
whole thing, and will, I think, go away after
the Bladders. A heavy gale blew from the SW
during the evening, a most awkward direction
for us.

Wednesday & Thursday. May 19 & 20
Blowing a hard gale both days. We are
tacking and turning between the ice and
Spitzbergen. We can make out the Windward
in the lulls, sometimes ahead, sometimes astern.
Sea running very high, and sky as dark
as possible. Took a sea aboard on Wednesday,
giving the Watch a fine ducking. My old foe,
Toothache, has it seems followed me all the
way from Scotland, and been hiding about the
ship the whole voyage. Yesterday it came out
from its concealment and said "Oh, mine
enemy, and have I found thee?", whereupon
it seized hold of me by one of my incisors
and twinged it so, that my whole face is
distorted today. (Addison). On Thursday we saw
the wild bleak coast of Spitzbergen breaking

through the rifts after the Storm. A great line of
huge black perpendicular crags running up to
several thousand feet, as black as coal but
all seamed with lines of snow. A horrible
looking place. We were thinking of running
in and anchoring in King's Bay, but the
chart was mislaid. Toothache.

Friday May 21ˢᵗ

Spitzbergen still in sight about 50 miles
to the North East. A complete lull in the wind
but the sky very dark, and a heavy swell on,
from which we think we will have a change
of wind. Windward went South, and in the
evening we saw the Eclipse coming up in the
distance. As we had not seen her for a month
we were anxious to know what she had done.
The Captain boarded her and after three hours came
back with the news that she had been down to
Iceland and had managed to capture 32 Bottle-
-nose whales, a very large take. They yield on an
average a ton apiece, and as Captain David had
also as many young and old seals as we he
has managed to beat us so far. 90 Tons I believe

he has got. This wind has done us terrible injury by packing all the ice up close, and destroying all the bights or bays in it in which whales are usually found. However we must just keep up our peckers, and hope for the best.

Saturday May 22nd

A heavy swell all day. I come of age today. Rather a funny sort of place to do it in, only 600 miles or so from the North Pole.

The "Hope"

"The Seven

through the rifts of the Storm. A great line of huge black perpendicular crags running up to several thousand feet, as black as coal but all seamed with lines of snow. A horrible looking place. We were thinking of running in and anchoring in King's Bay, but the chart was mislaid. Toothache.

Friday May 21st

Spitzbergen still in sight about 50 miles to the North East. A complete lull in the wind but the sky very dark, and a heavy swell on, from which we think we will have a change of wind. Windward went South, and in the evening we saw the Eclipse coming up in the distance. As we had not seen her for a month we were anxious to know what she had done. The Captain boarded her and after three hours came back with the news that she had been down to Iceland and had managed to capture 32 Bottle-nose whales, a very large take. They yield on an average a ton apiece, and as Captain David had also as many young and old seals as we he has managed to beat us so far. 90 Tons I believe

he has got. Th
by packing a
all the brights
usually four
our pickers,

Saturday m
A heavy swell
Rather a fun
600 miles or

wind has done us terrible injury
the ice up close, and destroying
bays in it in which whales are
   However we must just keep up
d hope for the best.

22nd

ll day. I come of age today.
sort of place to do it in, only
from the North Pole.

The "Hope" in a gale off Spitzbergen.
May 20th 1880.
"The Seven Ice mountains" to the Left.

Had rather a doleful evening on my birthday, as I was very seedy from some reason or another in the evening. The Captain was very kind and made me bolt two enormous mustard emetics which made me feel as if I had swallowed mount Vesuvius, but did me a lot of good. Eclipse sailing near us all day. Ice is sadly damaged.

Sunday may 23d

Plum Duff day again – a fine day, the swell all gone. Sailing in to the West again down the tight ice. The Captain and I have been making most villainous parodies of Jean Ingelow's "Sparrows Build"

"when Sparrows build & the s

"When 'Bergies' Build their Greenland nest,
    my spirit groans and pines
For I know there are Seals in the Nor' Nor' West
    But its time to 'coil our lines.'
Far down in the South the 'Bladders' lie
    But the Devil a one near me,
And the 'Unis' are sticking their horns on high
    as they plunge & play in the sea

But oh the Whale, and the right, right, whale!
And the Whale we all love so
Is there never a 'bight' in the Greenland 'tight'
Where a whale has room to blow.
                    12 foot whale can

Thou didst set thy foot on the ship and fare,
  To that sad and lonely shore,
        cold
Thou wert sad for the seals were all skin & hair,
  And they came from Labrador.
And 'Meg' he came and scared away
  Some Twenty mile at the least,
And how could we tell where the 'Flappies' lay,
  With a great bay flaw to the east.
                                    Chorus.

We shall never again sail back in May,
  As we oft before have done
Or take four thous and 'Young' in a day
  And go home with two hundred ton,
We shall never be full with seals alone
  For all our work and toil,
But we'll never say die while whales yield bone,
  And 'Bladders' give up their oil
                                    Chorus.

Monday May 24th

Another fine day. We are going to have a little luck at last I hope. 6.P.m. no. we are not through. We are certainly awfully unlucky this year. A strong wind has set in from the East and is packing up a nice little bight which was forming, and playing the deuce with our prospects. Colin says we have a Jonah aboard. Eclipse near us. Got our harpoon guns stuck up.

Tuesday May 25

Worser and Worserer. Wind still blowing from the East and murdering the ice. All hands disgusted. Eclipse set sail for the South but seemed to think better of it and came back again. Horrible!

Wednesday May 26th

A fine day but the ice is ruined. Wormed our way through it as best we could. I was smoking my afternoon pipe on the quarterdeck when there was a cry of ' A Bear - close to the Ship' Captain was at the masthead and sang out at once to 'Lower away the Quarter Boat.' I ran down for my shooting iron and succeeded in getting a seat in the boat. I could see the bear - a great brute - looking quite tawny against the white snow, and running very fast in a direction parallel to the ship. Then he crouched down in a hole of water about a couple of hundred yards off, and hid with just his nose above surface. Mathieson was harpooner of the boat and we pulled off, but had to make a bit of a circuit to get through the ice. We lost sight of him; and when we saw him again he was standing with his fore paws on the top of a hummock, and his head in the air, staring at us and sniffing. We were within shot then but we thought he would let us get nearer, so we bent to our oars. again But some associations connected with boats seemed to dawn on his obtuse intellect, for he suddenly got off the hummock and we lost sight of him. Then we saw the signal hoisted for the boat to come aboard, and spied Bruin travelling over the ice at a great rate, and a long way off, so we had to give it up as a bad Job. Wind still Easterly.

Average for the Old Sealing

Buchan	14	+	87	+	75	+	18	+	20	=	214	
Colin	10	+	85	+	51	+	8	+	7	=	161	
Hutton	9	+	112	+	11	+	6	+	12	=	150	
Carner	10	+	61	+	42	+	11	+	2	=	126	
McKenzie	2	+	55	+	41	+	8	+	5	=	111	(mob)
Rennie	13	+	68	+	26	+	7	+	2	=	116	
Mathieson	10	+	47	+	10	+	11	+	2	=	80	
Kane	1	+	36	+	20	+	2	+	13	=	72	

Scramble 55

Young Sealing 141

Total. 1216 Old Seals

The Bear we did not shoot
may 26th 1880

Thursday May 27th

Ice began to close round us fast in the morning,
and we had to steam our way out to the open sea
as best we could, to save ourselves from being beset
or nipped. Had a difficult job to get out. Eclipse
kept in our wake. Captain went on board her to
dinner and stayed till about eight. I drew, slept,
played draughts and boxed while he was away. We
are going off to 80° North Latitude to see what is
up there, right up to the Northern Barrier in fact.
The terror of the seas up here is an animal which
is called the Swordfish, but is not a swordfish
at all. It is one of the whale tribe with a long
snout like a mackerel, and great pointed
teeth the whole length of its jaws. It attains the
length of 25 feet, and is distinguished by a
high curved dorsal fin. It feeds on the largest
sharks, on seals and on whales. Yule of the
'Esquimeaux' took six whales the other year in
the straits, which actually came and cowered
under the ship for protection, because one of
these monsters was in the vicinity. The captain
tells me that he was in the Crow's Nest one day

when he saw a great hubbub a head of the ship. On examination with the glass he made out, that it was an enormous sea elephant which was sitting on a piece of ice very little larger than itself. In the water round it were half a dozen of these blood thirsty fish, which were striking the poor creature with their long fins, trying to knock him off his perch when they would have made short work of it him. As the ship came up the Captain says he never shall forget the look which the poor seal cast towards it with its big eyes, and suddenly taking an enormous bound off the bit of ice, it squattered along the Surface of the water, and took such a leap towards the ship's side, that its head was above the taffrail, and it very nearly gained the deck. A boat was lowered when the great 12 foot creature climbed into it, and was knocked on the head. Balls had to be fired into the fish to keep them from attacking the boat they were so riled at the disappearance of their prey.

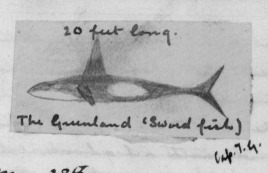

20 feet long.

The Greenland 'Swordfish.)

(cp. T.G.)

Friday May 28th

Steaming North and North East all day, in company with the "Eclipse". It is clear we are not going to have any whales in May, and we can only hope for the best in June. Thick fog in the evening and we had to blow our steam whistle and fire guns for several hours before we could find the "Eclipse" which was also screaming loudly. Took a bang at some loons on the ice at a long range with my rifle, since signals were the order of the day, but allthough I hit the piece and knocked the snow all over them I slew none.

Saturday May 29th

The less said about Saturday the better. Let Saturday sink into oblivion. Nothing doing. Fog in the evening. Lat 79.10 North at noon. Played cards in evening.

The 'Loon' or Lesser Auk.

Sunday May 30th

Captain David came aboard in the morning
and expressed great dissatisfaction at the state of
ice, in fact he said he had never seen it worse. W.
Walker came afterwards with a log book for me
which the Captain very kindly sent. In the morn
we espied two objects swimming near the ice, which
the Captain made out to be two Ground Seals, a
rare variety, nearly as large as Bladder noses.
We lowered away a boat and after an exciting c
and an exhibition of bad shooting on the part of
harpooner we nailed them both. They were a fem
and young one, the former about 8 ft. 6. By the
way talking of Bladder noses Colin killed one on
which measured 14 feet long. It charged the boat
and nearly bit the harpoon gun in two. We hope
we may have a turn in the luck now after this
small capture. By the way one of the most
interesting things in Arctic Zoology was the captur
last year of a large Albatross by Capt David,
in 80° North Latitude. Where did the bird come
from? It looks as if the temperature of the Pole
was semitropical

Monday May 31st

    Dreamed about whales in the Caledonian Canal and how frightened we were lest some of the barges or horses would scare them. There were 17 of them, all bottlenoses under a bridge. A very curious dream. The wind this morning is WNW which is excellent, and the water is of a greenish hue which is excellenter.

S.

a Capture.

[S]unday May 30th

Captain David came aboard in the morning
[an]d expressed great dissatisfaction at the state of the
[ice], in fact he said he had never seen it worse. D[r]
[Wa]lker came afterwards with a log book for me
[wh]ich the captain very kindly sent. In the morning
[we] espied two objects swimming near the ice, which
[the] Captain made out to be two Ground Seals, a
[rar]e variety, nearly as large as Bladder noses.
[We] lowered away a boat and after an exciting chas[e,]
[an]d an exhibition of bad shooting on the part of the
[har]pooner we nailed them both. They were a female
[an]d young one, the former about 8 ft. 6. By the
[wa]y talking of Bladdernoses Colin killed one onc[e]
[wh]ich measured 14 feet long. It charged the boat
[an]d nearly bit the harpoon gun in two. We hope
[we] may have a turn in the luck now after this
[sm]all capture. By the way one of the most
[inter]esting things in Arctic Zoology was the Capture
[las]t year of a large Albatross by Capt David,
[in] 80° North Latitude. Where did the breed come
[fro]m? It looks as if the temperature of the Pole
[wa]s semitropical

Monday May 31st

Dreamed abo[ut]
and how frighten[ed]
or horses would [be]
all bottlenoses [in a]
dream. The wind
is excellent, and
which is excellent[ly]

a co[ast]

whales in the Caledonian Canal
we were lost some of the barges
ue them. There were 17 of them,
a bridge. A very curious
morning is W N W which
water is of a greenish hue

Swordfish in pursuit of a school of Seals

By the way I haven't half exhausted my curious
dream. While we were away killing the whales
under the canal bridge I heard it strike two
o'clock, and it suddenly came into my head
that my final professional was to have begun
at one. Horrified I abandoned the whales and
rushed to the University. The Janitor refused me
admission to the examination room, and after
a desperate hand to hand struggle he ejected me
Even then I did not wake but dreamed that
some one handed me out a paper to see what
the questions were like. There were four questions
but I forget the two middle ones. The first was
"Where is the water ten miles deep near Berlin"
The last was headed NAVIGATION, and the question
was this word for word "If a man and his
wife and a horse was in a boat, how could the
wife get the man and the horse out of the boat
without swamping it?" I grumbled very much
at these questions, and said it was not fair to
introduce Navigation into a medical Examination
Then I determined to send the paper to Captain
Gray and get him to answer it, and then at last

I woke up. Certainly the most connected dream, as well as the most vivid I ever had.

This evening our foretop yard came down with a run owing to the breaking of a shackle, and smashed the halliards. We put up a spare spar and made all right again within four hours, a fine bit of seamanship. Captain has gone up to the Nest and I am writing this before the cheery cabin fire. I hear the hammering on deck as they do up the broken yard, and just outside the door the Steward is remarking in an really first class tenor that ' at midnight on the Sea-heas - Her Bright Smile haunts me still'. It seems to haunt him at midnight, and then he employs the odd 23 hours in commenting upon the fact." Captain David was on board in the evening, and lent me a pamphlet on whales. I was experimenting on the 'maukies' in the evening. I took 4 pieces of bread and soaked them, one in Strychnine, one in Carbolic Acid, one in Sulphate of Zinc, and one in Turpentine. Then I threw them over to the birds to see which would work quickest, but to my horror and old patriarch stepped forward and swallowed the whole four

pieces,

and strange to say he didn't seem a bit the worse

Tuesday June 1st

I trust that we may have better luck this month,
than last. We can see the Northern Barrier along
the whole horizon. Eclipse is getting up steam
and I suppose we are going to have a look at
the bight in Lat 78° out of which we were
driven, and then we will run away to Scoresby
Sound on the Liverpool Coast if nothing turns
up. Water is full of animalculae and olive
green in Colour,

              Balæna Mysticetus!
                   Balæna Mysticetus!
                If we were animalculæ
                      You wouldn't take long to eat us.
Captain says he has seen whales spouting so
thick that it looked like the Smoke of a large town
a very good description. By the way the water rose 8
degrees yesterday from which we think we are in the
Gulf Stream. Passed a piece of a fir tree floating
in the water, It has come many thousands of miles,
drifting down the Obi or Yenesei rivers in Siberia
and so into the Arctic seas by the N W current.

Saw
~~got~~ two bladdernoses in the evening but only got
one on account of bad shooting. Hoped to get away
and shoot roaches in the evening but there weren't
any. Buchan shot four in the morning.

~~Tuesday~~ __Wednesday June 2.__

Plying West and South under canvas. Captain
suffering from Ablubberomnia— Very cold, as cold
as it was in April. My hair is coming out and I
am getting prematurely aged. Read a good story
that a doctor was buried in the middle of a large
churchyard, and a professional brother suggested as
an epitaph 'Si monumentum quæris, circumspice'
Very witty, I think. It is very disheartening to be kept
off the whaling banks like this by the ice. As the Steward
says 'it makes a lad inclined to Jump up, and never
come down again'. Sydney Smith said of Jeffrey
'His body is too small to cover his mind. Jeffrey's
intellect is always indecently exposed'. Very clever too.
Saw the marks of a large bear in the evening, also
a Bladdernose in the water. Things look rather more
hopeful this evening, as we have made considerable
way to the Westward, and are close to the whalebanks.

Thursday June 3d

Very cold again, a great hoar frost is on.
Strong wind from the North. We are on the
most promising place we have seen yet
and if the wind holds we ought to catch
some of the minnows we are after. Came
on a fog in the evening. I have my cod line over
board baited with pork but have not had a bite.
About 60 sail of Russians come to Spitzbergen
this month to hunt cod, so there must be some
knocking about. Had the bag net out tonight
and towed it to see if there was any food.
Brought up a most beautiful Clio or Sea
Snail, a couple of inches long, looking like
some weird little fairy. I have stuck him
in a pickle bottle and christened him "John
Thomas." I hope he will live, we have put
some butter and pork into his house. Saw
a good many narwhals knocking about,
one very large one, almost snow white and
quite 15 feet, rushed ahead past the Stern,
making the peculiar grunt they give when they
rise. Also saw some beautiful medusae.

———

Friday June 4th

John Thomas is in an awful passion. We left the pickle bottle far from the fire, and as there are 11 degrees of frost it froze up and John has caught cold. He is sitting in a corner with his tail in his mouth, just as a sulky baby sticks its thumb into its potato box. I have drawn John's attention to the butter & Pork and he took a hurried breakfast, but seems to have business of importance down at the bottom of the bottle. His thinking perhaps of

'where his Rude Shell by the Gulf Stream lay,
There were his little Sea Snails all at play,
There their Amæboid mother, He their sire
Butchered to make a Whale's holiday.

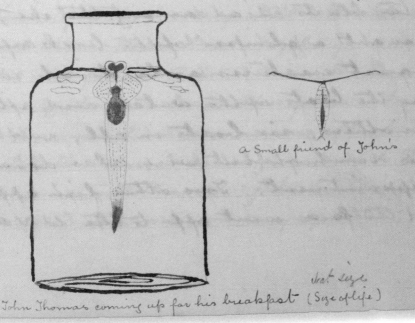

John Thomas coming up for his breakfast (Size of life)

a Small friend of John's

Nat size
(Size of life)

129

Just after one o'clock I was standing on deck talking to Andrew Hutton about the general bleakness of affairs. He is one of our best harpooners. I happened to ask "By the way, Andrew, when a man does see a whale I suppose he never sings out 'There she blows' as put in books." He said "Oh they cry 'There's a fish' or any thing that comes into their head, but there's Colin going up to the masthead, so I must go on the bridge'. Up he went on the Bridge and the moment he got there he bellowed out 'There's a fish'! There was a rush for the boats by the watch but the Captain put a stop to it "Do things quietly", he said "and man the boats when I give the word". We could see two blasts ahead among the ice, and I caught a glimpse of the back of one of the great creatures as he dived. We lowered away the boats of the watch and afterwards four others, six boats in all, and our hopes ran high but were alas doomed to disappointment. Two other fish appeared, and the four went off to the W N W. The

boats kept after them for four hours, but it was
no go, something (heavy ice) seemed to have scared
them. However it is something to be on their
trail. We see a flaw water ahead and hope we
are going to have a fine time of it in there.
The Eclipse was also after two fish but lost them
owing to the unskilfulness of their second mate
who lost his post in consequence.

Saturday June 6th

John is well and hearty. Saw a great many
narwhals today, but none of what we want.
Kept a lookout on the bridge from breakfast
to dinner. We saw a large Sea Elephant on
the ice about noon, and Andrew Hutton &
I went away and shot it. About 9 feet
long and very fat. We opened its stomach
and found it contained a very large assortment of Cuttlefish. Captain went aboard
the Eclipse in the evening. The guano here
is blood red, and has a curious effect. Plenty
of birds about. Wind coming round to the
South. It is a most exciting business; the
tension on the nerves is very great

Sunday June 6th

John was up before me and took a heavy breakfast. He is now gyrating round the top of his bottle surveying his new kingdom apparently and meditating a map. I put him in a bucket every evening where he wanders fancy free for an hour or two. Wind is round to S W I am glad to say, it was S S W yesterday. We may see fish any moment now; the water is a peculiar dark grey green. I thought I smelled a fish yesterday from the deck. You can often smell the queasy smell long before you see them. Aaron our Shetland boy the son of old Peter the prophet was in the crew of the boat that visited the Eclipse yesterday when he came back I heard him go straight to his father and begin with "Father, Peter Shene's been dreaming!" (Peter Shene is the rival prophet of the Eclipse 'Ay, boy. What?
' Peter Shene's had a dream, Father.'

'and what did he tream, joy?'

'He treamed he saw them killing cows on poard the Hope.'

'Oh a goot tream, joy, a goot tream. That means that the Hope will have the first fish. A very goot tream.'

So we still have some hopes.

Saw a large Cuttle fish under the surface, and a good many medusae and Clios. About 3 PM Word came that the Eclipse Boats were away. They were several hours after their fish but finally they were recalled by the hoisting of the bucket. About 6 PM Adam Carner saw a blast a long way off from the masthead. Four boats were sent off in pursuit, but failed even to catch a glimpse of the whale. Jock Buchan, who by the way started in his shirt and trousers just as he tumbled out of bed, nailed a Narwhal or Sea Unicorn about 13 feet long, with a horn of 2 feet. The Harpoon cut its throat most beautifully. It was towed by the four boats and hoisted aboard. Beautifully speckled with Black and Grey.

after flinching it we opened the Stomach which we found to be full of a very large shrimp, which I take to be the 'mountebank' shrimp, and with bits of cuttlefish. It had two distinct sets of para-sites upon it, one like a long thin worm in the drum of its ear. The other and like at the root of the horn.

Two very rare ducks were seen behind the ship this evening. The Captain went off himself in a boat and nailed them both with a right and left barrel. No one on board has seen the species before. They have a yellowish beak with an orange callosity stretching up in a curve from the base of the beak towards the eyes. They are rather larger than our ducks, dark brown in the head, white on the neck, dark brown in the back, lighter silvery brown on the breast. All the plumage very soft & delicate

Towing home the Narwhal

The Narwhal itself

Monday June 7th

No fish seen today though the Captain thought
he had a glimpse of one in the evening. I
went aboard the Eclipse after tea to get some
arsenical soap to preserve our ducks with.
Captain David says he thinks they are King
Eider ducks, a very rare bird. Captain David
came back with me and stayed an hour.
He was after three fish yesterday but got none.
I caught a petrel by flinging a lead with a
bit of string attached over its head, when
the string wrapped round it and I hauled
it in. It looked confoundedly astonished. I
let it away again. Wind North & North East.
Blowing hard in the morning

Tuesday June 8th.

Steamed a bit in the morning. Sun
shining brightly. Secured the ship by an
anchor to a piece of floating ice, and
whistled for a change of wind. Sent
away three expeditions after Narwhals
but without any success. Went away
birdshooting but only got two shots

killing a Roach and maiming a loon.
Peppered a Flaw Rat but it got away. Grant
saw the steps of an Arctic fox in the ice.
Had a pleasant day on the whole. Captain
says if I will load my own cartridges I may
blaze away until all is blue. Made sail
again in the evening. Played 'Nap' in the
engine Room. Almost dead calm.

Wednesday June 9th

We were forced to come out towards the open sea
again today on account of changes in the ice.
Eclipse and we moored ourselves on to one
piece of ice in the evening. Captain David &
Dr Walker came aboard us about ten Pm
and stayed until two. They shot a very
large bear upon the ice today. It was sitting
munching away at the head of a narwhal
which it had dragged on to the ice, while a
great shark was wiring into the tail which
hung over into the water. How the bear got a
narwhal onto the ice is a mystery. Went away
at Two o'clock in the morning to shoot birds
They were very scarce however and I was

only enabled to get a roach and 5 Snowbirds
Saw a large Bladdernose but were unable to
get a shot at it. Came back at 4 P.m.

Thursday June 10th

Still trying hard to get in to where we know
the whales are lying. Made some progress
under steam and then anchored with the
"Eclipse" to an iceberg. Shot a Kittiewake and
a Loon off the deck, and then got two more
loons while picking up the first. Amused
myself in the afternoon by catching petrels
by flinging a lead over the heads of them,
and warping the string round their wings,
something like the South American 'Bolas'.
By the way when I shot a roach the other
day a great Maulie seized it the moment
it fell and regardless of the shouts of the
boats crew, and my frantic howls proceeded
to bear it away, but I shied a boats
stretcher at it and scared it off.

Maulie Stealing our Roach

He was a right thinking and high minded Clio,
distinguished among his brother sea-snails for
his mental activity as well as for physical
perfection. He never looked down upon his
smaller associates because they were protozoa
while he could fairly lay claim to belong to the

End of Volume II. Log of Hope.

high family of the Echinodermata or Annulosa.
He never taunted them with their want of a
water vascular system, nor did he parade his
own double chain of Ganglia. He was a modest
and unassuming blob of protoplasm, and
could get through more fat pork in a day
than many an animal of far higher pretensions.
His parents were both swallowed by a whale
in his infancy, so that what education he had
was due entirely to his own industry and
observation. He has gone the way of all flesh
so peace be to his molecules.

Zoological List of Whaling Voyage

<u>INVERTEBRATA</u>

I Protozoa.

    Any number in Whale's food

II Infusoria

    Rice food.

III Annulosa

Common Louse (On a Shetlander)	Clio Helicina
Shrimp (Common)	Horn louse of Narwhal
Clio Borealis (John Thomas)	Ear louse of Narwhal
Shrimp 'Mountebank'.	Whale Louse (Ocina)

IV Echinodermata

    Medusa gileus   78.40 N.

    Medusa —— ?   78.40 N.

    Flask Shaped Medusa   78.5 N.

V. Mollusca

    Sepia ——   78.40 N.

VERTEBRATA

I. AVES

Arctic Petrel or "Maufie" (Procellaria Glacialis) —

Foolish Guillemot or "Loon" (Colymbus Troile) —

Roach (Arca Alle) —

Doveca (Colymbus Grylle) —

Burgomaster (Larus Glaucus) —

Kittiwake (Larus Rissa) —

Snowbird (Larus Eburneus) —

Snow Bunting (Emberiza Nivalis) —

Redpoll. (Fringilla Linaria) 75 N.

Puffin or "Tammy Norie". (Alca Artica) 78 N.

Boatswain (Larus Crepidatus) 78·12 N

Iceland hawk (Falco Icelandicus) 73·40 N

Great white owl. (Stryx) 71 N.

Great Tern or Sea Swallow (Sterna hirundo) 78·18 N

Brent Geese. (Anas Bernicla) 78 N

Eider duck (Anas mollissima) —

Sea Gull (Larus Communis)

Shag. (Lerwick).

Duck (Calva?). Very rare. King Eider. 78·50 N.

Arctic Starling 78·6 N.

Sandpiper 75·30 N.

Arctic Gull. 69° N

141

PISCES

Flaw fish (Rather like a whiting) 78.40 N.

Silver Fish 78.12 N.

Herring. 69°N.

Squalus Greenlandicus or Greenland Shark

MAMMALIA.

Horse saddle Seal. (Phoca Vitulina)

Bladdernosed Seal

Flaw Rat.

Ground Seal 79 N.

Walrus. (Trichecus Rosmarus) 77.30 N.

White faced Seal.

Fresh Water Seal. (78 N)

Orca Gladiator. The Greenland Swordfish. Lat 69 N.

Bottle nosed Whale (Delphinus Weductor) 68 N.

Razor back Whale (Balæna Phypalis)

Narwhal (Monodon Monoceros)

Right Whale (Balæna mysticetus).

Balæna Musculus (Hunchback whale) 68° N.

Ursus Maritimus or Polar bear

Canis Lagopus or Arctic Fox

              Additional birds seen on passage.

Solan Goose.

Stien Chuck

Stormy Petrel.

Black Back Gull.

Sparrow Hawk

Mallet

"Log of the Steam Ship 'Hope'.

Greenland whaling & sealing.

Summer 1880.

vol. III.

Friday ~~~~~~~~~~~~

~~~~~~~~~~ made a few miles in the right direction. The Eclipse shot two bears this morning. About one o'clock a fish came up near the Eclipse but was not captured. We made fast to an iceberg in the evening. Caught a curious fish today, the first I have seen in Greenland. It looked rather like a whiting, but was not one. Jack Williamson one of our hands got a terrible blow from the wheel ~~today~~. It exposed the bone of his skull for about 5 inches. Stitched it up and sent him to bed Steward the boatsteerer and I were walking on the ice in the evening and both distinctly saw the blast of a fish about half a mile off among the pack ice. It could not however be reached by boats

Saturday June 12th.

The ice is shutting rather than opening. Hope deferred maketh the heart sick. Men shot a bear off the side about eight o'clock. I was asleep and so missed the fun. Stomach was full of seal oil but he was very thirsty [illegible] the [illegible] shots. Went aboard the Eclipse at dinner time with the Captain. Had a pleasant feed and chat. Captain David seems far from despairing. Strong wind from W and SW— ought to do us good. Ice began to close round us so rapidly that we had to steam out 30 miles or so to prevent being nipped or beset. Had a difficult job to get out as it was. The sea this morning was actually swarming with Narwhals.

Sunday June 13th

Got a fine opening towards the Westward and worked in again about as far as we came out, going W & SW. Saw nothing but one seal in the water the whole time. Need about 20 miles North to take us into whaling ground. 'Thou art so near and yet so far'. It does seem hard after our penetrating impenetrable packs, and leaving forty miles of shifting heavy ice between us and the sea, exposing ourselves to every danger of storm and flood, and putting ourselves in the way of losing our ship and ourselves, or of being beset and wintering out, that we shd be no better off than the half hearted beggars who shun the whole concern, and go South after small game. It is a shame and a sin, and can't last long. The Eclipse and ourselves are the last of two generations of daring Arctic seamen, the breed has deteriorated and we are the sole survivors of the men who used to harry Greenland from the 80 to the 72, and here we are stuck in the mud & helpless. It would make a saint swear.

Monday June 14th.

Thick fog in the morning. Blew foghorns but got no answer from the Eclipse. Jack Williamson, the man with the head is doing very well. Things look as bad as they can be and worse. I hope we will go and hunt bladdernoses instead of persevering at this. The whales are only 20 miles off but an impassable barrier of ice intervenes, and the wind is such as to pack the ice firmer together, rather than to open it out. We want wind from the W. NW or NWW and we are getting it all from the South. ο ποποι ποποι!   53 tons!   Such is life!

149

Tuesday June 15th

The only difference in the weather is that the fog is thicker and the wind more utterly odious and depraved. However we are at the bottom of our woes for nothing could make our situation worse, so 'there's an end on't' as old Sam Johnson used to say. Captain went aboard the Eclipse at dinner time. I do hope we'll go and slay bladdernoses or Bottle- -noses or any other animal who has some peculiarity about its nose, and carries blubber on its carcase. Browsed over Boswell all day.

Wednesday June 16th

The Eclipse lowered away after a whale about 8 AM and pursued it until noon, but did not get it. The fact is we are not upon the grounds, and any we see are stragglers on the march, and not stopping to feed. Wind Westerly, so far so good. Calm in the evening, sea looks like quicksilver, the whole place covered with Narwhals, great brutes 15 & 16 feet long. You hear their peculiar 'Sumph!' in every direction.

I saw one pass like a great flickering white Ghost underneath the keel. Reading "Tristram Shandy"; a coarse book but a very clever one.

Thursday June 17th

An eventful day - for the Eclipse at any rate. In the morning about 10 AM Colin saw a whale from the Crows Nest about five miles off while the Eclipse had her boats after another. We made sail and reached up towards Colin's fish but did not see it again until about 1 PM when it suddenly appeared within 50 yards of the ship, accompanied by another one. The two were gambling and frisking in the water like a couple of lambs. We lowered away four boats, Colin's, Carner's, Rennie's and Peter's which all pulled up for a piece of ice where the fish were likely to reappear. They came up there near Rennie's boat, but he unfortunately is not a man of much decision of character, and he hesitated to fire into the nearest fish for fear of scaring the other, which was turned eye on to him. The fish separated one disappearing and the other leading the boats a most exciting a most exciting chase to Windward. From the deck I could see its blast rising apparently just in front of the boats, and its great tail waving in the air, but our men could never get quite within shot of it. The Eclipse seeing the way the whale was heading came round that way and dropped two boats in front of it; The whale came up in front of one, the second mate's, and in a moment we had the mortification of seeing the boat's Jack flying, as a sign they were fast to our fish. It is hard to see a thousand pounds slip through your fin--gers so. They killed the fish during dinner and had it aboard before 8 PM.   Rennie got a fine wigging from the Captain when he came aboard.

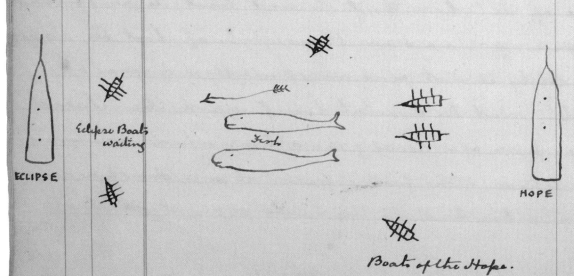

ECLIPSE   Eclipse Boats waiting   Fish   HOPE

Boats of the Hope.

After dinner we saw a large bear on a point of ice apparently in a great state of excitement, probably due to the smell of the whale's blood. I got off in the boat with Mathieson to kill it. We got out on the ice, a great flaw many miles across, with our rifles, and could see the bear poking about among some hummocks about 40 yards away from us. Suddenly he caught sight of us and came for us at a great speed, running across the pieces of ice towards us, and lifting up his fore feet as he ran in a very feline way. Mathieson and I were kneeling down on the ice, and I intended not to fire until the brute was right on the top of us. Mathieson however let blaze when it was about 15 yards off and just grazed its head. It turned and began trotting away from us, and as it only presented its stern I was compelled to put my bullet into that. It was wounded but went went off across the ice at a great rate, and we never saw it more.

Saw a 'Boatswain' Gull today. Boo Row in the mate's berth in the evening

Friday June 18th

Eclipse struck another fish during the night and had it aboard before breakfast. Lucky dogs!. Buchan shot a fine bear and two cubs during the night. By the way my bear of yesterday when it had escaped some distance got up on its hind legs on the top of a hummock like a dancing bear, to have a good look at us. I had no idea they did that in a state of nature. Cruised about all day in search of blubber but found none. Our boats and the Eclipse were after one fish at ought but they never got a start. Ice is closing round us and we are cut off from the sea, so that unless there comes a change of wind we may easily be beset. We are all very melancholy.

Our Bear.

Saturday June 19th

Calm as a fish pond, water like quicksilver. A good many Narwhals about. The ice is remaining stationary or thereabouts. No fish seen today at all. A Shark was seen to come up alongside and nail a

markie out of the water. We wouldn't mind being
unsuccessful if others shared the same fate, but it
is maddening that the Eclipse should make £3000
while we have not made a penny. Our Captain
is as good a fisherman as ever came to Greenland,
there are no two opinions on board on that point,
quite as good as his brother David, but somehow
the luck seems to be with the others. They have seen
from first to last about 14 fish to our 5.

Sunday June 20th

A large fish was seen during breakfast, but
after a short pursuit it got among pack ice
where it was impossible to follow it. It was very
nearly within reach once. This is terrible to see
fish and not to get them. No man who has not
experienced it can imagine the intense excitement
of whale fishing. The rarity of the animal, the
difficulty attending any approach to its haunts,
its extreme value, its strength, sagacity and size
all give it a charm. A large bear was shot
during the night

Ice closed round us during the day but relaxed towards
evening. Captain David came aboard during the
day, and our Captain went and had supper with
him. One of his harpooners was attacked in his boat
by a bear the other day when he had no rifle with
him, but he banged the hard wad of the harpoon gun
through it, which was ingenious. That was nothing
however to what one of our harpooners did a few
years ago, which would be incredible if I did
not know it to be true. Buchan was sent to shoot
a bear and two cubs on the ice, but they took to
the water before he reached them. He passed the
noose of a rope over the head of each, as they
swam and snarled at him, and tied the ends
of the three ropes to his thwarts. All the oars were
then run in except the steer oar, and Buchan
standing in the bows and banging them on the
head with a boat hook whenever they offered to
turn, guided the boat right back to the ship
the bears towing it the whole way. The Stewart who
saw it, says the roaring could be heard a mile

off. Some hopes for tomorrow

Monday June 21st

Hopes not realized as usual. We are shut up in a small hole of water with nothing but great ice fields as far as we can see. If it goes on shutting up we will be beset and our chances of any success ruined. Colin bad with a Sore throat. No fish about. Caught a beautiful Sea Lemon yesterday floating on the surface but it died shortly after being brought on board. Saw a very curious sight at midnight, which you might come North a lifetime and never see. There were three distinct suns shining at the same time with equal brilliancy, and all begirt by beautiful rainbows, and with an inverted rainbow above the whole thing. A most wonderful spectacle.

A Family party
(Seen from the deck of the Eclipse).

Tuesday June 22nd

An utterly uneventful day. We are still cooped up in this hole of water. Caught a rare & indeed un-described medusa in the evening (Medusa Woilea Octostirpata). Misery and Desolation

Wednesday June 23d

Made our way out of our prison, by a most delicate and beautiful bit of manœuvring under steam. We came out about 60 miles among very heavy ice, the smallest piece of which could have crushed our ships like eggshells. Often we squeezed through between floes where the ships sides were grinding against the ice on each side. Steamed S and E. Eclipse went after a fish but never got a start. Glass falling rapidly

Thursday June 24

Captain and I were knocked up at 6 AM by the mate's thrusting his tawny head into the Cabin, singing out 'a Fish, sir', and disappearing up the Cabin stairs like a Lamplighter. When we got on deck the mate's and Peter's boats

were already on the seat of action where the fish
had been seen. They caught another glimpse of
it about a mile to leeward and pulled down
towards it, but lost sight of it again. Meanwhile
another very fine whale came up astern very
near the ship and Hutton's and Rennie's boats
were lowered away after it. The four boats

...ul day. We are still cooped up .
...aught a rare & indeed un-
...the evening (Medusa Wailea
...and Desolation

...of our prison, by a most
...bit of manœuvring under
...about 60 miles among very
...piece of which could have
...eggshells. Often we squeezed
...here the ships sides were
...ce on each side. Steamed
...t after a fish but never
...lling rapidly

...knocked up at 6 AM by
...tawny head into the Cabin,
...sir", and disappearing
...ke a Lamplighter. When
...mates and Peter's boats

Boats of the Eclipse a...

...were already on the seat of
...had been seen. They caught
...it about a mile to leeward
...towards it, but lost sight
...another very fine whale ca...
...near the ship and Hutton's
...were lowered away after it

Hopo in pursuit of two whales.   June 17th 1880.

tion where the fish
nother glimpse of
nd pulled down
it again. Meanwhile
up astern very
nd Rennie's boats
The four boats

pursued it for a couple of hours, when it began to blow extremely hard and a heavy sea arose so that some of the boats' head sheets were right under water. We had to get them on board, and let the whale alone. Blowing a very hard gale from the North East all day, 9 Wind Force.

Friday June 25th

Wind is still very strong though not so much so as before. Nothing seen during the day but a large Finner whale which is a bad sign. It is of no use to us, and it drives the right whale away from its feeding grounds. Played Nap in the evening. Wind only a fresh breeze now, have begun to steam to the North after the Eclipse

Saturday June 26th

Things looked dark enough all day, but suddenly took a turn for the better. Nothing had been seen all day and I had gone down to the cabin about 10 o'clock when I heard a sort of bustle on deck. Then I heard the Captain's voice from the masthead

"Lower away the two waist boats!" I rushed into the mates' berth and gave the alarm, Colin was dressed but the second mate rushed on deck in his shirt with his trousers in his hand. When I got my head above the hatchway the very first thing I saw was the whale shooting its head out of the water and gambolling about at the other side of a large "sconce" piece of ice. It was a beautiful night, with hardly a ripple on the deep green water. In jumped the crews into their boats, and the officers of the watch looked that their guns were primed and ready, then they pushed off and the two long whale boats went crawling away on their wooden legs one to one side of the bit of ice, the other to the other. Carner had hardly got up to the ice when the whale came up again about forty yards in front of the boat, throwing almost its whole body out of the water, and making the foam fly. There was a chorus of "now, Adam — now's your chance!" from the line of eager watchers on the vessel's side. But Adam Carner, a grizzled and weatherbeaten harpooner knows better. The whale's small eye is turned towards him and the boat lies as motionless as the ice behind it. But now it has shifted, its tail is towards them – Pull, boys, pull! Out shoots the boat from the ice – will the Fish dive before he can get up to it? That is the question in every mind. He is nearing it, and it still lies motionless – nearer yet and nearer. Now he is standing up to his gun and has dropped his oar – "Three strokes, boys"! he says as he turns his quid in his cheek, and then there is a bang and a foaming of waters and a shouting, and then up goes the little red flag in Carner's boat and the whale line runs out merrily.

But the whale is far from taken because it is struck. The moment the Jack appeared in the boat there was a shout of 'a fall' on board, and down went other six boats to help the 'fast' boat and nail the whale on its reappearance. I got into the mate's boat and away we pulled. Of course the whale may come up anywhere within a radius of the line it has taken out, which may amount to three or four miles, so our seven boats had to spread out over a considerable area. Five minutes passed — ten — fifteen — twenty and after being away 25 minutes the brute came up between the second mate's and Rennie's boats, who fired into her and despatched her. She proved to be a small fish, about 40 feet long, with 4 ft 1 in of bone, worth between £200 and £300. We gave three cheers and towed her to the ship. She was covered with very large crab lice which accounted for her erratic conduct in the water. Had her flinched and stowed away by 3 o'clock AM. I went to the Crow's nest during the process to look out for another which I didn't see. Went to bed at 6 AM and got up at 12.

Adam Carner.

Sunday June 27th

Not a thing to be seen all day, but about 4 AM Colin saw a very large fish in the distance. Eclipse lowered away 2 boats as well as we, and after getting one start they lost scent of her. She seems to have been a tremendous brute.

Monday June 28th

Nothing

Tuesday June 29th

Master aboard the Eclipse last night until 2 AM. Lay to in a calm all day - nothing doing. Waiting for a chance of getting North but ice looks bad. Got away to shoot at midnight and came back at 4 PM. Got a Burgy, a Snowbird and five Loons. These Burgies I think are the biggest of Gulls after the Albatross. They usually are about 5 feet from tip to tip

Wednesday June 30th

Slept nearly all day after last nights exertion. Went aboard the Eclipse boat and had a great talking. Worked with a microscope aboard. Buchan shot a flaw Rat. Hulton skinned my birds ursday and Friday July 1 + 2.

Lying to in a thick fog as we have been ever since last
Sunday. Nothing to Chronicle except that Colin got a large
narwhal early on Thursday morning. It took out a whole
line (120 fathoms) and made a great fuss about being
killed. A Unicorn is worth about £10. The skin is
of considerable value. I have a very decent Arctic museum by this time including a lot of interesting things.
I have at present

1. An Esquimeaux pair of Sealskin Trowsers
2. An Iceland Falcon
3. My Sealing knife and Steel.
4. Bone of Bladdernose — Shot myself.
5. 2 Bones of Old Seals
6. 2 Foreflippers of Young Bladdernose
7. 2 Foreflippers of a Ground Seal
8. A Bear's head
9. Bristles of a Bladder
10. A Burgomaster
11. Drums of Whale's ears
12. 2 King Eider ducks

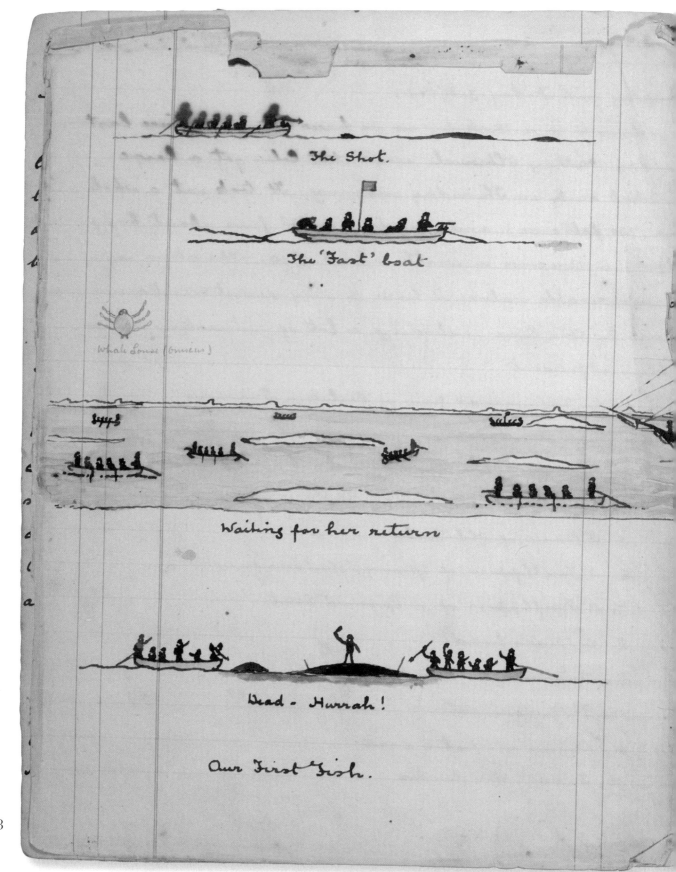

The Shot.

The 'Fast' boat

Whale Louse (oniscus)

Waiting for her return

Dead – Hurrah!

Our First Fish.

13. Bits of Lava found in King Eider duck.

14 (?) A Unicorn's horn.

Added 2 Esquemeaux pouches
a Kittiewake
a Bear's Claw

Saturday July 3d

It has cleared up and we are off to the happy hunting grounds. Sailed Nor' and Nor' West all day. Saw nothing but an extraordinarily small seal on the ice, about the size of a rabbit. It seemed as much amused at the appearance of the ship as we were at it. We are all despairing. The Steward stuffed my Ground Seal's flippers very nicely with sawdust.

Sunday July 4th

Sailed North and then South again. Everything looks bleak and discouraging. No trace of whales or even of whale's food. A Bladdernose was seen on the ice. A small bird something like a starling or thrush was flying round the ship. Saw a puffin. Have no heart to write much in the Log. Reading Morley's "Rise of the Dutch Republic"; a very fine history.

Monday July 5th

Monday J. Steamed into a flaw water, made fast in the evening. Saw several 'Finner' whales. Eclipse mistook

169

one for a fish and lowered away his boats, which however were promptly recalled. Nothing else of interest seen during the day. Got some delicious fresh water off the salt water ice

Tuesday July 6th

Dead calm. Sun beating down in a tropical manner though the temperature was only 36°. Tremendous glare from the ice flaws. Went aboard Eclipse in morning. Got away to shoot and nailed altogether 7 Loons, a roach, a Kittiwake, a Snowbird, and a flaw Rat. We had great fun securing the latter as our small shot did not suffice to kill it, and after a chase of atleast half an hour. we harpooned it with boat hooks when swimming under the water. We brought it aboard alive but the Captain humanely put it out of its misery. Got away again at night but found no game. A couple of Sea Swallows played round the ship. Saw several Finners. A very jolly day!

Wednesday July 7th

Steamed 20 or 30 miles South, and then on seeing indications of fish made sail. Captain David

came aboard in the evening with his ~~harpooner~~ Engineer.
and caught a rare shrimp. Feel very much
the better for yesterday's outing. Cooked Red
Herring for supper in a very scientific manner.

Thursday July 8th

Another memorable day. Sailed along the edge of a great
flaw among very blue water with the Eclipse ahead of us.
About 1 o'clock a whale, the first seen since Sunday
week came up close to Captain Davids ship; he lowered
away three boats after it and chased it until 4.15 Pm
when he succeeded in getting fast, and had her along-
-side by 8 Pm and flinched by midnight. We dodged
about hoping his fish would come in our direction
when we would have been Justified in securing it,
but about 4 o'clock the welcome shout came from the
masthead "there's another fish on the Lee Bow, Sir!"
Mathieson and Bob Cane lowered away after it, and
were soon lost sight of among the ice, while we
crowded along the side of the deck and waited. Then
a groan went up as a large Finner whale rose
near the ship, for Finners and "right" whales are

171

deadly enemies, and we were afraid our quarry would be scared. I went down to the Cabin to sooth my disappointment with a smoke, when I heard the Captain yell "A fall! A fall!" from the masthead, which is the signal that the fish is struck. Up we tumbled many of the men only half dressed, and away went five long green whale boats to the support of the 'fast' boat and it's companion. I got into Peter McKenzie's boat. We had hardly got clear of the ships side when the boat steerer announced that the fish was up, and was lashing out, fin and tail. Then we knew our work was cut out for us, for when a fish stays a very short time under water after being struck, it is reserving all its strength for a struggle with the boats. If the whale goes down and stops away half an hour it is generally so exhausted on returning to the surface that it falls an easy prey. The boats pulled up and Hutton and Carner fired into her and got fast. We were the next boat up and pulled on to

fishes head, where we lanced her deep in the neck.
She gave a sort of a shudder and started off at a
great rate along the surface. Buchan pulled his boat
on to her head as she advanced, by which senseless
manoeuvre the prow of the boat was tilted up in the
air, and finally the whole boat landed on the
animal's back amid a shouting of men & snapping
of oars and Buchan roaring "Pull! Sweep! Back!
Hold Water! Pull! What the devil are you feared at!
I said to Peter "Stand by to pick them up!" but they
managed to shove the boat off without accident.
The beast now made for a flaw and got beneath
it, but soon reappeared when both Buchan and
Rennie fired into her. She went under again but
once more reappeared right among our three
boats and then the fun began. We pulled on to
her and in went our lances for five feet or so.
the three boats tried to keep well at the side of her
while she was always slewing round to bring her
~~tail~~ formidable tail to bear upon us. She nearly
had our boat over once by coming up underneath

173

it, but we managed to get it righted. Then we stood off from her while she went into her dying flurry, whipping the water into a foam, and then she slowly turned up on her back and died. We stood up in our boats and gave three hearty cheers. We towed her up to the ship and by 1 P.m had her aboard. She was a fine fish, each lamina of whale bone being 9 foot 6 inches, yielding about 12 tons of oil. It is worth quite £1000 and has secured our voyage from being a failure. A large and very ugly shark came up and superintended the process of flinching the fish in spite of numerous knives passed through its body. I asked the Captain to let the Steward and me go off in a boat and harpoon it, but he refused.

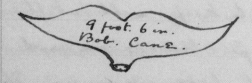

9 foot. 6 in.
Bob. Cane.

Whale

Buchan's Boat on the top of the fish

Friday July 9th.

Nothing doing. Everybody in a state of reaction after yesterday's capture. The Captain says Bob Cave managed the affair very well. Several Finner whales seen during the day. Beautiful sunshine.

managed to get it righted. Then we
from her while she went into her dying
hipping the water into a foam, and then
y turned up on her back and died. We
in our boats and gave three hearty
e towed her up to the ship and by 1 P.m
aboard. She was a fine fish, each
f whale bone being 9 foot 6 inches,
about 12 tons of oil. It is worth quite
has secured our voyage from being
. A large and very ugly shark
and superintended the process of
the fish in spite of numerous knives
trough its body. I asked the Captain
Stewart and me go off in a boat
oow it, but he refused.

9 foot 6 in.
Bob. Cane.

Friday July 9
nothing doing.
yesterday's cap
-aged the affa
seen during the

whale dragging 2 fast boats through water
July 8th 1880

...chan's Boat on the top of the fish

...erybody in a state of reaction after
... The Captain says Bob Care m--
...very well. Several Finner whales
...day. Beautiful sunshine.

It is a curious fact that the last whale the Eclipse captured only had one eye, and our friend of yesterday was also restricted to the same meagre allowance. The socket was perfectly empty. It may be that there is a breed of one eyed Greenland whales

Saturday July 10.

We have made a mistake, I think, in heading North again. The South seems to me a greasier locality. Had the boat away after a Bladder which we did not get. Played Euchre four hours in the evening in the Engine room. Query. Who did Adam & Eve's children marry?

Sunday July 11.

Got up late, and would have liked to have got up later, which is a sad moral state to be in. Eclipse got a Bladder in the morning. Steamed to the Eastward with the Eclipse in the evening, by which proceeding we scared a whale. Saw many Finners. About seven Pm a steamer was reported about 20 miles to the Eastward. This is the first ship we have seen since the beginning of May. We steamed out and soon recognized it as the new discovery yacht of Leigh Smith's the "Eira". He is going to try for the Pole if the ice is favourable, which it won't, and in any case to explore Franz-Joseph Land and shoot deer. He is a private gentleman, a bachelor with £8000 a year, and has taken Spitzbergen to himself as a wife. When our ships came up we saluted the little Eira with ensigns and three cheers, which they returned. His men are in naval reserve uniform, officers in gold lace. The Captain went aboard her, while their Doctor, Niel, the Photographer, the Engineer and 2 mates all boarded us. The Captain came back about 10 PM and he and I with the Eira's Photographer & Doctor made a night of it on champagne and sherry. We had tinned salmon at 5 AM and turned in at 6.30.

Monday July 12.
Anchored to a flaw with the Eclipse and Eira. Unshipped our rudder which was damaged by

ice. By the way we got our home news up to June. 13th from the Eira. Got no letter from Edinburgh but a very pleasant and cheerful one from Tottie. Surprised to hear that the Liberals have got in, disgusted also. Invited aboard Eclipse to meet Lugh Smith and Gang at dinner. Had Mock Turtle Soup, fresh Roast Beef with potatoes, French beans and Sauce, Arrowroot pudding and pan- -cakes with preserves, winding up on wine & cigars. A very respectable whaler's feed. Went aboard "Eira" afterwards, She is beautifully fitted up aft. Had more cigars and Champagne. Got aboard at 12, after being photographed in a group. They came up by Jan Meyen and saw millions of bladdernoses in the 72.30, I hope we may come down for them.

Tuesday July 13.
Steamed 20 miles South and stopped short, I don't know why. I fancy we might fill our ship now if we went straight down for those Bladders but we must go at once. We are vacillating here too

much, I think, however it is for the Captain to decide.
The success of our voyage depends on these few days.
its our last chance of making a hit. Eclipse
chased a bear and killed it in the water close
to our ship. Left the Erna in the morning wishing them
all success. A pleasant ship and a pleasant crew. She
is black with a line of gold, about 200 tons burden
and 50 horse power engines. I think I should like to be
going out in her, although the prospect of seeing home
again is pleasant. They left their letters with us. Fog in
the evening

Wednesday July 14

Steamed and Sailed South and Sou' West. Eclipse
had their boats away in the evening, but it was only a
Finner which they mistook for a right whale. Foggy
nearly all day. No news of any sort. Read our
papers ~~all day~~

Thursday July 15.

Another uneventful day. Lounged about & smoked.
Absolutely nothing to do. Very thick and foggy and all
that is reprehensible. Saw a small scene of Goethe's

'Faust' which I am reading which I think is as vivid and weird as anything I ever read, far more gruesome than Shakespeare's witches.

Night - - - An Open Plain

Faust. Mephistopholes rushing past on black Horses

Faust – What are these hovering round the Ravenstone?

Meph – I know not what they're shaping & Preparing

Faust – They wave up – wave down – They bend – They stoop.

Meph. A band of witches

Faust. They sprinkle and Charm

Meph. On! On!

That is very awful I think.

Friday July 16.

Still foggy. Eclipse had four boats away during the night, but without success. We do not know yet if he really saw a fish. Boarded him in the evening and learned that it most certainly was a fish, and that they very nearly secured her. They got near enough to touch her with a boathook as she swam under water. Captain David still seems to be very sanguine. Some of our stores are running short, we got some potatoes however from the Erik ~~hower~~. Stayed till 2 AM on the Eclipse. Got some more papers. Many seals seen during the day in the water.

Saturday July 17.

Absolutely nothing to do or to be done. It has been thick fog now for nearly a week. Steamed about 20 miles S and E. Captain David came aboard at night. We intend now to try the Liverpool coast right down in West Greenland near the land Lat 73° N. Many heavy fish have been taken there late in the season by Capt David, notably in '69 when he took 13, striking the first on the 16th of July, and the last on the 4th of August. I remember I used to think that when a whaler saw a whale they always got it, as a matter of fact the average is about one fish in 20 attempts.

Sunday July 18.

A little clearer today, not very much. Strong SSW breeze changing to a gale in the evening. Blew very hard all day, and all night. Dodged about under the lea of ice flaws to escape its fury. I wonder if Leigh Smith's vessel is caught in it. By the way I was photographed among a distinguished group on the quarterdeck of her, but as I was smoking a cigar during the operation I am afraid I'll be rather misty

Monday July 19.
Blowing a gale all day. Nothing to do and we did it.

Tuesday July 20.
Cleared up a little and we did a good day's work steaming among great icefields about 40 miles S & W. If it keeps clear we may do something yet. There is an enormous accumulation of ice this year round the land, more than has ever been seen. We are 240 miles from it now, and the fields are almost continuous. I'm afraid we won't get in

Wednesday July 21.
Thick again, this fog is paralyzing — We are groping in the dark. Anchored to a flaw in the evening and the Captain and I went aboard the Eclipse. Had

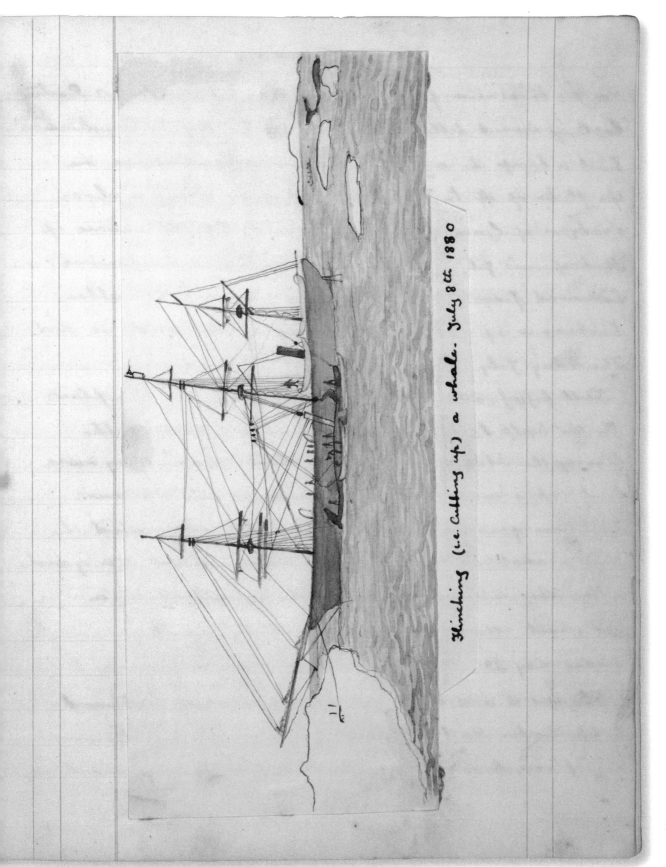

Flinching (i.e. Cutting up) a whale. July 8th 1880

185

Nox Ambrosiana from 8 to 2 AM. The Late Mr. Proctor. Captain David tells me of a fish he captured which had a lump the size and shape of a beehive on the fluke of its tail. He entered into a critical analysis of Goethe's Faust comparing it with some of Shakespeare's plays, and showing where the former borrowed from the latter, so we are not altogether barbarous up here.

Thursday July 22.

Still foggy and we continue anchored to the flaw. In the half deck in the morning discussing the loss of the Atalanta. Saw 2 'Boatswains', very rare birds at a considerable distance over the flaw and was going to hunt them but they absconded. Got a shot at a flaw rat's head about 50 yards off in the water, and blew it clean off with a rifle ball. Unfortunately the body sank.

Friday July 23.

Steamed S and SW as it became clearer. Continued to ply under sail all night in the same direction among very heavy ice fields. Wind coming round to the Westward.

Saturday July 24.

Steamed SW again all day. Went through some ice
that would have made Sir George Nares and the whole
Arctic Committee turn up the whites of their eyes. Looking back it seemed solid as far as the horizon, and
you could hardly conceive that two ships had
formed their way through it. We have one or two
faint hearted ones aboard who have the terrors; it is
not the going in it is the going out again, they say
and we only have a fortnight's provisions left.
If we got beset we should certainly have to go on
uncommon short commons. We are leaving 200
miles of heavy ice between us and the sea.

Sunday July 25
A very clear day with occasional fogs. Steamed
40 miles West. Made sail in the evening. Saw a
great number of 'boatswains'. We have been exulting
rather during the last few days as we have been getting
on to Westward very well, but our way seems to be
barred now by an immense chain of flaws, which

we hope to circumvent.

Monday July 26

Sailed West and South West. Made our Longitude
6¼° W, and our Latitude 73-56 N. Captain went
aboard Eclipse in evening. Water swarming with
food but no animal life to be seen save 9 maukies
and a school of Phoca Vitulina. Wrote a POM,
about a Meershaum Pipe

It lies within its leather case,
  As it has lain in years gone bye,
Trusty friends and Comrades true,
  Are that old Meershaum pipe and I.

For it was young when I was young
  And many a Jovial reckless night,
We students drank, and smoked and sung,
  While yet my meershaum pipe was white ~~young~~

And it was hardly brown before,
  From home and friends I first did part,

But bound for Russia's hostile shore,
    I bore my meershaum next my heart.

And there upon the bloodstained ground,
    Where many came and few went back,
With death and pestilence around
    Twas there I smoked my meershaum black.

And when the day our Colonel died,
    We charged and took the Malakoff,
A Russian bullet grazed my side,
    And shot my meershaum's amber off.

But I am grizzled now and bent,
    Death's sickles near – His crop is ripe,
I fear him not but wait content,
    I wait and smoke my meershaum pipe
                                    ALD

Tuesday July 27th

Flying under sail about SSW. Latitude at noon gave us 73.29 N. A large Finner whale, the first we have seen for some time came up below the quarter boats. It seems to be a disputed point whether they are a good or a bad sign, the majority affect the latter opinion, but Captain David Gray throws his very weighty verdict on the minority. From my own experience I should say that the presence of Finners is not by any means a bad sign.

Blew a fresh breeze in the evening, ice moving at a great rate
Spent some time in the half deck. 'Eric' built a house as a
depot in Davis Straits. On returning one season they found a
polar bear lying asleep in one of the beds on the top of the blankets.
Reading Maury's 'physical Geography of the Sea. He explains
the weed of the Sargasso sea (on the triangle between Cape de
Verdes, Azores and Canaries) by saying it is the centre of
the whirl of the Gulf stream, as when you whirl the water
in a basin, you find floating corks at centre. He also
remarks that railway trains always run off the line to
the right hand side whether going North or South

      Wednesday July 28.
Another disagreeable day. Blowing hard from the South East,
which is about the worst possible direction. This is the
longest interval we have ever had. The ship has not drawn
blood since July 8th, except a flaw rat I shot. Blew from
Eastward in the evening. As thick as pea soup and ice
closing upon us rapidly. We have hopes that there is the
open sea to the South of us from the fact that seals are
coming through from the South. I thought too there was a
swell from the same direction which would settle the
                    question.

Tuesday July 27th

Plying under sail about SSW. Latitude at noon gave us 73. 29 N. A large Finner whale, the first we have seen for some time came up below the quarter boats. It seems to be a disputed point whether they are a good or a bad sign, the majority affect the latter opinion, but Captain David Gray throws his very weighty verdict on the minority. From my own experience I should say that the presence of Finners is not by any means a bad sign.

'Flinching a Fish'. Sketched by Capt David
S.S. Eclipse

Blew a fresh breeze in the evening, ice moving at a great rate. Spent some time in the half deck. 'Eric' built a house as a depot in Davis Straits. On returning one season they found a polar bear lying asleep in one of the beds on the top of the blankets. Reading Maury's 'Physical Geography of the Sea'. He explains the weed of the Sargasso sea (in the triangle between Cape de Verdes, Azores and Canaries) by saying it is the centre of the whirl of the Gulf stream, as when you whirl the water in a basin, you find floating corks at centre. He also remarks that railway trains always run off the line to ___ ___ de whether going North or South

28.

___ day. Blowing hard from the South East, ___ worst possible direction. This is the ___ have ever had. The ship has not drawn ___ except a flaw rat I shot. Blew from ___ ning. As thick as pea soup and ice ___ pidly. We have hopes that there is the ___ th of us from the fact that seals are ___ n the South. I thought too there was a ___ e direction which would settle the
question.

A 'Right' and 'Left' among the Loons.

A very anxious and disagreable night for us all, blowing hard, thick fog and ice everywhere. Captain and I could not turn in till 4 AM.

Thursday July 29

Horrible contemptible pusillanimous thickness over all. made fast to a flaw, and waited for better days. Went on a Journey over the ice, accompanied by our newfoundland Sampson. Were out of sight of the ships and had great fun. Came across a most extraordinary natural Snow house, about 12 feet high, shaped like a beehive with a door and a fine room inside in which I sat. Travelled a considerable distance, and would have gone to the pole but my matches ran short and I couldn't get a smoke. Got a long shot at a Boatswain but missed him. Steamed SE when it cleared, but as it grew thick again we had to anchor once more. Eclipse shot at a Bladder but missed it. Got a curious fungus on the ice.

Natural Ice house. Lat 73.15. Long 6 W.

Gin and Tobacco at night

Friday July 30

Suffered for the Gin and Tobacco. A most lovely day 72.52. N. Jan Meyen bearing SW about 100 miles and not visible. Steaming SSE at 6 knots. Took no dinner but went to the masthead in preference enjoying a pipe and the welcome sunshine. Fell in with one or two small bladdernose seals of which we shot two, one fell to my rifle, the other was the object of the worst exhibition of shooting I ever had the misfortune to witness, I fired my only cartridge at a long range and missed, whereupon two harpooners took the Job in hand, and fired 3 shots each, or 7 shots in all before the unfortunate seal dropped its head.

Saturday July 31st

Out in the open sea pitching and tossing like Billy and with her head WSW bound for the Bottle Nose Bank. It is very problematical whether we will get any of the creatures, as I suppose they shift their ground like all other animals in these regions, and because Captain David got them there in April is no reason why we should see them again in August. No ice in sight. I shall never again see the great Greenland floes, never again see the land where I have smoked so many pensive pipes, where I have pursued the wily cetacean, and shot the malignant Bladdernose. Who says thou art cold and inhospitable, my poor ice fields? I have known you in calm and in storm and I say you are genial and kindly. There is a quaint grim humour in your bobbing bergs with their fantastic shapes. Your floes are virgin and pure even when engaged in the unsolicited 'Nip'. Yes, thou art virgin, and drawest but too often the modest veil of Fog over thy charms.

I can apostrophize the icefields, but hang the word will I say in favour of Spitzbergen, the Jotunheim of the Scandinavian mythology which I saw in a gale and left in a gale, a barren rugged upheaval of a place. Sailed West and Sou'West all day. It fell calm in the evening and we lay in a long rolling swell our sails flapping and a thick mist around.

Sunday August 1st

Eclipse out of sight – has probably been steaming in the fog all night. Steamed W and SW through calm water and thick mist. We hope we may find Bottle-nose whales about 80 miles SE of Jan Meyen, and from there to Langaness in Iceland. Keeping up our spirits. Saw some drift wood today. Hove a bottle overboard in the evening with our Longtitude and Latitude and a request to publish where it was picked up. Bottlenose fishing has never as yet been atall developed, several ships have tried it in a half hearted way and failed. The Jan Meyen got 9 in 6 weeks which did not pay them, Capt David this year got 32 in a month which did pay him. Fell in with very greasy water tonight, with a strong smell of

herrings and swarming with clios, I caught about 100 of
the shelled variety. One would think the Bottlenoses
would be near such tempting Grub. Heard a Finner
whale blowing away in the mist like an empty beerbarrel.
Lat 70.59. Long 0° 15 E. Passed 2 dead maulies and another
bit of drift wood from Siberia. Several more 'finners' seen.

Monday August 2nd

Sea calm and hardly any wind. The top of Mount Beerenberg
is in sight bearing WNW about 80 miles. Saw several
Puffins, seaswallows and eider ducks, birds only seen in
the vicinity of land. About two o'clock four Bottlenose
whales, two old and two young came in sight and
two boats were lowered away in pursuit. They made
straight for Cane's boat but when within shot they dived
and though we pursued them two hours we never got
another chance. About 5 o'clock two more came up and
Colin was sent after them but they disappeared. The Eclipse
is in sight and had his boats away also without success.
They are funny looking brutes in the water, with high
dorsal fins like finner whales. They are worth about £60
each. Quite warm now, have all our flannels off.

Bottle Nose whale in water

Tuesday August 3d

    Things don't look as well this morning as there is more wind and not so many birds or food in the water.
Sailing Westward. Nothing seen during the day

Wednesday August 4th

    Came into better ground this morning, there being very many birds and much grease on the water. Watched the Bósun gulls, who are very bad fishers, chasing the poor old Kittiewakes until they disgorged their last meal, which the bullies devour in its semidigested condition. Sea was swarming with cetaceans about noon which we lowered away 2 boats for thinking they were bottlenoses but they proved to be young finner whales, worthless brutes and so powerful that they would run out all our lines, so the boats were recalled. Captain shot a "Boatswain". Saw many Eider ducks. Several swordfish also seen. One of them was chasing a "Finner" whale round the Eclipse. The poor brute was springing right out of the water and making an awful bobbery. Carner put a rifle bullet into one young one about 40 feet long, which went away in a great hurry to tell its ma what they had been doing to it. This sea from Jan Meyen to Iceland might be called the Feather sea. The surface is literally covered with feathers in many parts. This Bottlenosing is an awful spree.

'Hope' in a calm among Cetaceans. Aug 4th 1880.

Was called up about 11 PM by the Captain to see a marvellous sight. Never hope to see anything like it again. The sea was simply alive with great hunch back whales, rather a rare variety, you could have thrown a biscuit on to 200 of them, and as far as you could see there was nothing but spoutings and great tails in the air. Some were blowing under the bowsprit, sending the water on to the forecastle, and exciting our newfoundland tremendously. They are 60-80 feet long, and have extraordinary heads with a hanging pouch like a toad's from their under jaw. They yield about 3 tons of very inferior oil, and are hard to capture so that they are not worth pursuing. We lowered away a boat and fired an old loose harpoon into one which went away with a great splash. They differ from finner whales in being white under fins and tail. Some of them gave a peculiar whistle when they blew, which you could hear a couple of miles off

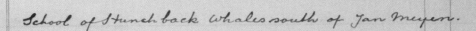

School of Hunchback Whales south of Jan Meyen.

Thursday August 5th

Nothing seen today. A stiff breeze arose towards evening and pitched and tossed us about confoundedly. We think the Eclipse has gone home. Steering SW

Friday August 6th

Gave it up as a bad Job and turned our head ESE
for Shetland. Dense fog and rain with very little wind.
Utterly beastly weather. We are all dejected at having to turn
home with so scanty a cargo, but what can we do? We've
ransacked the country and taken all we could get, but
this is an exceptionably unfavourable year owing to the
severity of last winter which has extended the Greenland
Ice far to the Eastward, and locked the fishes feeding ground
inside an impenetrable barrier.   Here is our whole game bag
for the season according to my reckoning

    2 Greenland Whales
    2400 Young Seals
    1200 Old Seals
    5 Polar bears
    2 Narwhals
    12 Bladder noses.
    3 Flaw rats
    1 Iceland Falcon
    2 Ground Seals.
    2 King Eider ducks.

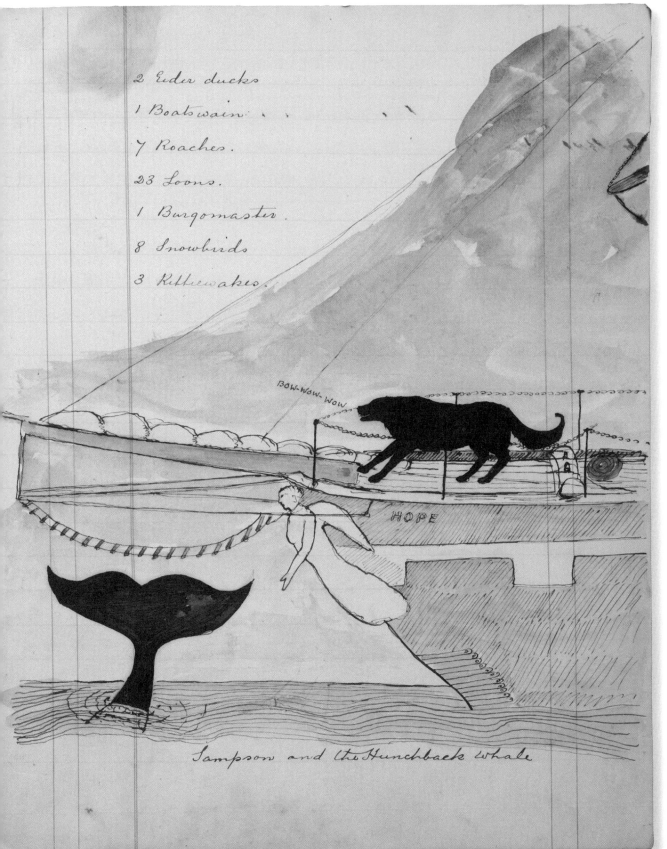

2 Eider ducks

1 Boatswain.

7 Roaches.

23 Loons.

1 Burgomaster.

8 Snowbirds

3 Kittiewakes.

BOW-WOW-WOW

HOPE

Sampson and the Hunchback Whale

Saturday August 7th

Groping homeward under steam and sail in such a thick fog that we can hardly see the water from the side of the ship. Took in the two Funnel boats. We have not got our reckoning now for several days, and as we have been dodging about zig zag after these bottlenoses, our dead reckoning is very uncertain. It isn't nice to be steaming along in the North sea in a fog with Iceland and the Faroe Islands knocking about in front of us. Several Puffins and other land birds seen.

Sunday August 8th.

Cleared up a little although it was raining nearly all day. Had a mackerel line over all evening but got nothing. Sighted land about 8 PM which proved to be the North end of Faroe island. A nice Job if we had come on it in the dark. Saw a Schooner running North about midnight, probably bound for Iceland from Denmark. Men busy drying our whale lines

Monday August 9th

A beautiful clear day with a blue sky and a bright sun. Wind from the NE a good strong breeze

before which we are flying homeward with all
sail set, and the bright green waves hissing and
foaming from her bows. No mackerel again.
Ship all covered with whale lines drying. Expect to
make the land late tonight. Saw a Solan Goose
and a little bird called a Stien chuck, also some
stormy petrels. The kittiewakes down here are a
smaller breed, I think, than those further north.
All hands on the lookout for land

     Tuesday August 10th

Up at 8 AM to see the land bearing WSW on the Starboard
bow. Half a gale blowing and the old Hope steaming
away into a head sea like Billy. Hence the feebleness
of my hand writing. The green grass on shore looks
very cool and refreshing to me after nearly 6 months
never seeing it, but the houses look revolting. I hate
the vulgar hum of men and would like to be back at
the Floes again

    "There is society where none intrudes
      Upon the sea, and music in its roar!"
Passed the skerry light, and came down to Lerwick but did

not get into the harbour as we are in a hurry to catch the tide at Peterhead, so there goes all my letters, papers and everything else. A girl was seen at the lighthouse waving a handkerchief and all hands were called to look at her. The first woman we have seen for half a year. Our Shetland crew were landed in four of our boats and gave 3 cheers for the old ship as they pushed off, which were returned by the men left. Lighthouse keeper came off with last week's weekly Scotsman by which we learn of the defeat in Afghanistan – Terrible news. Also that the Victor has 150 tons the dirty skunk. Took our boats aboard and went off for Peterhead full pelt. Fitful head and Sumburgh light twinkling away astern like a star. Herring fishing seems to be a success. Saw a large Grampus.

Wednesday August 11th

Dead calm and the sun awfully awful. Saw Rathay head at 4 PM. The sea black with fishing boats. Hurrah for home! Pilot boat came off at 6 PM and we lay off for high water at 4 in the morning. hundreds and hundreds of herring boats around us. Crew getting

on their shore logs.   Well here we are at the end of the
log of the Hope, which has been kept through calm and
through storm, through failure and success; every
day I have religiously Jotted down my impressions and
anything that struck me as curious, and have tried
to draw what I have seen.   So there's an

End of the Log of the SS Hope.

Our Illustrations

Fresh Meat.

Freemason's Hag

Rhamna Stacks

A Peterhead whaler

Sealing Costume.

The Hope among loose ice.

A family of Seals.

Seal Knife

Bear's Foot mark.

Ships among the Seals

Seal Club

Sketches at Young Sealing

Milne's Funeral

Our Hawk

Saturday's night at sea

A Snap Shot

Our Evening Exercise

All Hands over the bows

Plan of Seal fishing

Effects of refraction

Five Bulls at 100 yards

Vol II

Old Sealing

Making off Seal's blubber

Not sold but Given away

Poking the mob Boat

Harpoon Gun

Hope off Spitzbergen in a gale

The Bear we did *not* shoot.

Greenland Sword fish

The Lesser Auk (Loom)

Swordfish chasing Seals

A Capture

John Thomas & his friend

Towing the Narwhal home

A narwhal

Maulie stealing our roach

Vol III.

Capture by Eclipse's boats

Our bear

Bear and Shark devouring Narwhal.

Boats after 2 fish.

Our First Fish

Whale and 2 fast Boats

Buchan's Boat

Ship flinching a whale (Capt y Gray)

    do   do   do   (Capt David Gray)

A Right and Left among Looms.

Natural Ice house

Bottle Nose Whale.

Hope among Cetaceans

Hunchback Whales.

Sampson and the Fish

213

Game Bag of the Hope. (Continued)

July	1	1 Narwhal.
	2	
	3	
	4	
	5	
	6	1 Hlaw Rat 7 Loons 1 Roach 1 Kittiewake. 2 Snowbirds
	7	
	8	A Greenland Whale.
	9	
	10	
	11	
	12	
	13	
	14	
	15	
	16	
	17	
	18	
	19	
	20	
	21	

22			
23			
24			
25			
26			
27			
28			
29			
30			
31.	2 Bladdernose seals.		
August			
1			
2			
3			
4.	a Boatswain		
5.	2 Eider ducks		

My Own Gamebag.

Young Seals and Young Bladders xxxxxxxxxx· xx xxxxxxxxx· xxxx

xxxx· xxxxxxxx

Old Seals.   xxxxxxxxxx· xxxxxxxxxx· xxxxxxxx

Bladdernoses xxx

Loons   xxxxxxxxxx·

Roaches xxx

Maulies x

Snowbirds xx

Kittiewakes xxxx

Haw Rats xx

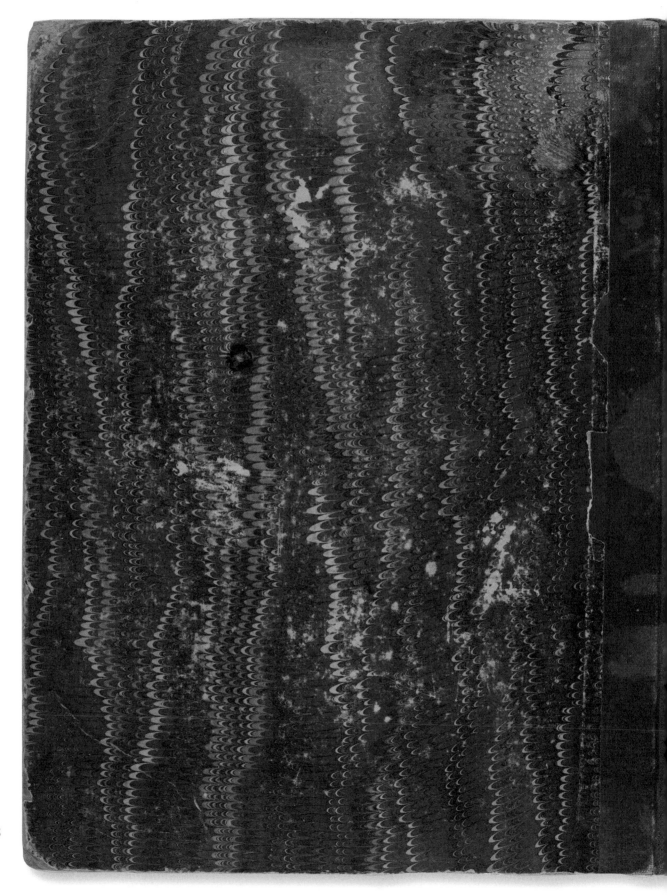

Annotated transcript
of Arthur Conan Doyle's
diary of his voyage

Transcribed and annotated by
Jon Lellenberg and Daniel Stashower

LOG OF THE S.S. *HOPE* 1880

Greenland whale and seal fishing

Saturday February 28

Sailed at 2 o'clock amid a great crowd and greater cheering.[1] The *Windward*
Captain Murray[2] went out in front of us, their Captain bellowing "port" and
"starboard" like a Bull of Bashan.[3] We set about it in a quieter and more
business-like way. We are as clean as a gentleman's yacht, all shining brass and
snow white decks.[4] Saw a young lady that I was introduced to but whose name
I did not catch waving a handkerchief from the end of the pier. Took off my hat
from the *Hope*'s quarterdeck though I don't know her from Eve. Rather rough
outside and the glass falling rapidly. Beat about the bay for several hours and had
dinner with champagne in honour of Baxter[5] and grandees on board. Pilot boat
came and fetched them all off at last, together with an unfortunate stowaway
who tried to conceal himself in the tween decks.[6] Sailed for Shetland in a rough
wind, glass going down like oysters.[7] As long as I stick on deck I'll do.

Sunday March 1st[8]

Got into Lerwick at 7.30 PM. Deuced lucky for us as a gale is rising and if we
hadn't made the land we might have lost boats and bulwarks. We were uneasy

---

1. And avoiding a common whalers' superstition against setting forth upon a Friday.

2. Captain Alexander Murray (1838–1894), for many years master of the screw bark *Windward* out of Peterhead. In 1893 the *Windward* was the last whaler to sail from there, captained by then by David Gray, elder brother of the *Hope*'s skipper.

3. Psalm 22:12 (King James Bible): "Many bulls have compassed me; strong bulls of Bashan have beset me round. They opened their mouths against me as a ravening and a roaring lion." In his poem "I Wonder What It Feels Like to be Drowned," Robert Graves writes of "a monstrous roar like thunder, / A bull-of-Bashan sound. / The seas run high and the boats split asunder...."

4. "Very unlike my idea of a whaler," Conan Doyle admitted in *Memories and Adventures* (1924).

5. William Baxter, owner of the *Windward*. "The men who financed and managed the Peterhead whaling fleet were for the most part merchants residing in the town," said a retrospective about "The Peterhead Whalers" in *The Scotsman* for November 19, 1902: "Shrewd men they were, cautious, but not afraid to lay out large sums of money when they had confidence in the captains and crews who manned their vessels. Not a few of them made considerable fortunes in the days when the fishing was at its best." Another was Robert Kidd, the *Hope*'s elderly owner of record through the Hope Seal and Whale Fishing Company.

6. The number of ships was the lowest in some years, according to the *Shetland Times* of March 6, 1880, and the number of seamen recruited in Lerwick commensurately low – some 244, nineteen of whom signed aboard the *Hope*, joining its Peterhead crew members. "Mostly old hands are being shipped," said the newspaper; many other aspirants did not find berths.

7. A lowered reading on the ship's barometer, or glass, indicated the approach of foul weather.

8. An error on Conan Doyle's part: 1880 was a leap year, and that Sunday, the 29th of February.

about it, but we sighted the Bressay light[9] about 5.30. Captain's very pleased we got in before the *Windward*, though they had 5 hours start.

Monday March 1st

Blowing a hurricane. *Windward* got in at 2 AM only just in time. The whole harbour is one sheet of foam. Feel very comfortable aboard. Have a snug little cabin. Telegraph gone wrong between this and Peterhead. Pokey hole.

Tuesday March 2nd

Glass down at 28.375. Captain has never seen it so low.[10] Blowing like Billy outside. Made out the hosiery list.[11] Tait on religion and atheism. He is our Shetland agent, not half such a fool as he looks.[12]

Wednesday March 3rd

Fine day. Glass still very low. Went on shore with the Captain after breakfast. Enlisted our Shetland hands. Fearful rush and row in Tait's small office. *Jan Mayen*[13] & *Victor* came in. Murray of the *Windward* seems a decent fellow. Captain [Gray] and I were going from Tait's shop when a drunken Shetlander got hold of him. "Capn, I'm (hic!) goin' with you. Oh such a voyage, Captain, such a voyage as never was landed! (hic!) Three hundred and fifty tons, sir, I've brought the luck with me." Gray turned back in the back room, and seemed annoyed. I said "I'll turn him out if you like, Captain." He said "Ah, I know fine you'd like a smack at him, Doctor. I would mysel but it wouldna do." We had locked the door of the back room when there was an apparition of a hand and arm through the smoked glass window which formed the upper half of the door. Bang! Crash! Wood and glass came rattling into the room, and we saw our indomitable Shetlander, with his hands cut and bleeding, looking through the hole. "Wood and iron won't keep me from you, Captain Gray. Go I will." The Captain coolly smoked his pipe the whole time and never moved from the stove. The man was carried off, kicking & thumping to the gaol, I suppose, though the

---

9. The Bressay Lighthouse two-and-a-half miles southeast of Lerwick.

10. "The lowest reading I can remember in all my ocean wanderings," wrote Conan Doyle forty-three years later. ("Whaling in the Arctic Ocean," in *Memories and Adventures*.)

11. "The surgeon's job was the easiest in the ship," says Alexander R. Buchan in *The Peterhead Whaling Trade* (Peterhead: Buchan Field Club, 1999), p. 56: "Apart from those times when he was compelled to apply his scant medical knowledge … his most important task was as a clerk," and this hosiery list is an example in Conan Doyle's case.

12. George Reid Tait "carried on an extensive and successful business as draper, clothier and shipping agent," says Thomas Manson's *Lerwick During the Last Half Century* (Lerwick: T. & J. Manson, Shetland News Offices, 1923), p. 89: "Mr. Tait had considerable interests in whale and seal vessels, for several of which he was agent, and during the seasons when the men were being shipped his shop was a very busy one" – as Conan Doyle notes in his next entry.

13. A Peterhead ship named for the volcanic Arctic island west of the North Cape discovered in the 17th century by Dutch and English whalers, now part of the Kingdom of Norway.

infirmary would have been a far better place. If ever I saw D T[14] that was it. Had Murray of *Windward* and Tait to dinner, talk of masonry, whaling &c. Dispute with Murray about efficacy of drugs. Plenty of good wine going. Finished the evening with the Captain very pleasantly.[15] By the way another stowaway turned up today, a wretched looking animal. The Captain was frightening him at first by telling him he'd have to go back, but he finally signed the Articles.[16]

Thursday March 4th

Gave out tobacco in the morning.[17] Slept forenoon. Went ashore in the evening. Went first with second mate and Stewart[18] to the Queen's[19] and had something short as he calls it. Then went to Mrs Brown's[20] and lost sight of them. Had a very hospitable reception there. Told me to make their home my home. Went down to the Commercial[21] where an F&E[22] was going on. Heard some good songs and sang Jack's Yarn.[23] Chat with Captain about Prince Jerome,[24] &c.

---

14. *Delirium tremens.* Robert S. Katz, M.D., a consulting pathologist with a special interest in A. Conan Doyle, points out a contradiction in the latter's diagnosis: if the Shetlander was drunk at the time, he would not have suffered simultaneously from *delirium tremens*, "which by definition represents an alcohol withdrawal syndrome; one can't develop an alcohol withdrawal syndrome if the patient is actively drinking." What seems beyond dispute is that the man was the worse for drink, whether or not actually imbibing at that moment. "His behavior," remarks Dr. Katz, "sounds rather more like someone who was just plain drunk."

15. Conan Doyle found his chief duty "to be the companion of the captain" because, observes Gavin Sutherland in *The Whaling Years: Peterhead, 1788–1893* (Centre for Scottish Studies, University of Aberdeen, 1993), p. 27, "the social gap between a skipper and his crew could not have been wider." Class distinctions were preserved aboard, the officers messing together at dinner "over which the skipper cracks his jokes, spins his endless yarns, and we talk over the events of the day," reported Cdr. Albert Hastings Markham, RN, an observer aboard the whaler *Arctic* in his previously cited journal, p. 20. Below decks, of course, the scene was quite different, says Sutherland: "A persistent fog of tobacco smoke filled the crowded crew's quarters and gin flowed liberally. Shetland fiddles played as bawdy songs were bellowed out with passion." Many were in their teens and twenties, like Conan Doyle, who was in a position to take part in both sides of the life aboard, and did.

16. Articles of Agreement, defining seamen's duties and responsibilities aboard ship.

17. Another clerk-like task of his as the ship's surgeon.

18. A typical spelling of the word in Scotland. Conan Doyle never gives the steward's name in his log, but knew and remembered it: Jack Lamb, the sparring partner whose "e'e" Conan Doyle had blackened on his first day aboard ship. "I don't know whether Jack Lamb still lives," Conan Doyle wrote, recalling the episode in *The Strand Magazine*, January 1897, "but if he does I am sure that he remembers the incident." Jack Lamb, twenty-seven years old at the time of the voyage, *was* alive in 1897: "My dear old shipmate," he saluted Conan Doyle, "when I look at the *Strand Magazine*, your article, we look back seventeen years and it seems as if yesterday. The way you describe everything is something splendid. It makes me think I can hear every word spoken yet, and the way in which you have described me, every one of my fellow servants knows it is me. I have lost the whiskers but the frame and bandy legs [are] as usual." If Conan Doyle by now was a world-famous author, Lamb had come up in the world as well. He had left the sea in 1881, entered the Royal Household at Balmoral, and was now the Queen's Baker at Windsor.

19. Queen's Hotel, built in the early 1860s hard by Lerwick's harbour, still in business today.

20. Maybe a family home, but not a boarding house, according to Douglas Garden of the Shetland Library, for 1879 and 1882 local almanacs do not show one, nor a seamen's institute, in Lerwick conducted by a Mrs. Brown. It may have been a brothel, if Conan Doyle "lost sight" of his mates at that point. Buchan's *Peterhead Whaling Trade*, p. 64, says that in Lerwick "the 'Southern boys' made the most of the inns, the shops, the fortune-tellers, and the loose women of the islands in a last fling before the grim times to come." Robert Smith's *The Whale Hunters* (Edinburgh: John Donald Publishers, 1993), p. 51, states: "Lerwick, with its narrow, refuse-strewn streets and ill-fashioned shops, its whisky-dealers and prostitutes, had a murky reputation."

21. Today the Grand Hotel, at 149 Commercial Street overlooking Lerwick's harbour.

22. "free and easy": a gathering in a public house or other such place for drinking and singing.

23. A nautical song celebrating the Royal Navy's opposition to the slave trade in the 19th century, written by Louis Diehl (music) and Frederic Weatherly (lyrics; also for the song "Danny Boy").

24. Jerome Bonaparte (1784–1860), Napoleon's youngest brother.

Friday March 5th

Captain and I were invited to Tait's for dinner. Both thought it a horrid bore. Went to the Queen's and played billiards. Then toddled down to Tait's. Met Murray of the *Windward* and Galloway, the latter a small lawyer, insufferably conceited – hate the fellow.[25] Had a heavy weary dinner with very inferior champagne. Old Tait expressed great surprise at my saying I was RD's nephew[26] – the old cow, I found out afterwards that the Captain had just been telling him about it. He has a dog who has been taught to love the name of Napoleon, if you talk of shooting Napoleon he will make a dart at you, and probably leave with some things of yours in his mouth, muscles and clothes and things. Murray talked about putting three men under the ice, seeing ten men shot in a mob row, and several curious things. We got the boat at nine o'clock and were both delighted to get on board again, and stretch our legs quietly. Wind rising. Saw what the Captain says is a Roman camp, but I think it is a round Pictish tower.

Saturday March 6th

Raining and blowing hard. Did nothing all day. Colin McLean[27] and men went ashore in the evening and hailed for boat, which we had to give though it was rather rough. Began Boswell's life of Johnson.[28]

Sunday March 7th

Nothing doing except that the mail steamer *St Magnus* came in with a letter from home and one from Letty,[29] also a week's Scotsmen.[30] Satisfactory news. We shifted our berth the other day in the harbour and now lie apart from the other ships with the *Windward*. Colin the mate was at the Queen's last night among a lot of Dundeesmen who spoke of those two D-d Peterheadsmen who went and moped by

---

25. James K. Galloway, a Lerwick solicitor thirty-four years of age, who evidently patronized him.

26. Richard Doyle (1824–83), one of his London uncles, famous as an illustrator for *Punch* along with much other work, including a satirical series on "Manners and Customes of ye Englyshe."

27. The *Hope's* forty-year-old first mate who posed ashore as a cook's assistant since he lacked papers. Conan Doyle was greatly impressed by his leadership, calling him "an officer by natural selection, which is a higher title than that of a Board of Trade certificate."

28. The first great English biography, of Samuel Johnson, written by his friend James Boswell and published in 1791. Throughout his log, Conan Doyle notes what he is reading in his free time. Some books he clearly brought with him, but others were likely borrowed from the Captain and perhaps other crew members, some of whom possessed an erudition that surprised him.

29. Maybe Frances Letitia Foley, an older second cousin in County Waterford, Ireland, of whom Conan Doyle was fond. He wrote his mother in November 1880, several months after the voyage, asking for Letty's address as well as his sisters Annette's and Lottie's, and in 1889 Letty Foley was on a list of family and friends asked to inquire for a new novel of his at their local libraries in order to boost interest. See *Arthur Conan Doyle: A Life in Letters*, eds. Jon Lellenberg, Daniel Stashower and Charles Foley (Harper Press, 2007). Her father Nelson Trafalgar Foley, of Ballygally House, Lismore, had been head of the Foley clan there up to his death a few years earlier, and a cousin of Conan Doyle's mother, Mary Foley, who came from Lismore.

30. Referring to *The Scotsman*, Edinburgh's principal daily newspaper.

themselves. Colin got up and after proclaiming himself a "Hope" man ran amuck through the assembly knocking down a Dundee[31] doctor. He remarked to me this morning when I was giving him a pick me up "It's lucky I was sober, Doctor, or I might have got into a regular row." I wonder what Colin's idea of a regular row is. Lerwick is a dirty little town with very hospitable simple inhabitants. Main Street was designed by a man with a squint, builded on the lines of a corkscrew.[32] Noticed today that some of the ships in harbour flew Freemason flags,[33] Murray has the Royal arch [Ark] up, Compasses on a blue ground. [DRAWING OF PENNANT]

Fishermen sell cod here at 5/ a hundredweight, and have caught as much as 25 Cwt in a night. By the way the Engineer of the *Windward* got his two forefingers crushed in machinery yesterday and I had to go over before breakfast and dress them. Twenty sail of whalers in the bay.

Monday March 8th

Nothing like a quill pen for writing a journal with, but this is such a confoundedly bad one. Went ashore today and after knocking about some time went up to see a football match between Orkney and Shetland – play rather poor. Met Captains of *Jan Mayen* (Denchars), *Nova Zembla*,[34] and *Erik*, also a London man, Brown, doctor of the *Erik*. Six of us went down to the Queens after the match and started on bad whiskey and went on to coffee. Then Brown ordered a bottle of champagne, and Murray and I followed suit. Cigars and pipes. I think we all had quite enough liquor. Brown was wrecked in the *Ravenscraig*[35] last year. Says he is a very superior sort of shot.[36] Captain and I got home about half past nine.

31. Dundee is a rival port on Scotland's coast, and never a favourite of Conan Doyle's – calling it in 1900, in a letter to his mother, "an odious place with every disadvantage."

32. The town's native chronicler, Thomas Manson, agreed (*Lerwick During the Last Half Century*, p. 84): "Lerwick is a curiously-contrived town, there is no doubt about that, and its main street is a huge joke – as a street. As a maze it is perfect, however." If Conan Doyle did not think much of Lerwick, neither did another young ship's surgeon beholding it twenty years earlier, J.F. Taplin of the *Windward:* "it's very small having only 3000 inhabitants, the streets are very narrow, not room for a cart to pass, and from all I could see I do not think they can use anything but thin little poneys [sic].... having posted my letter, I left the shore for my ship with the idea that I should never care to see Lerwick any more." Dr. Taplin's diary is at the North East Folklore Archive's website in Aberdeenshire, Scotland.

33. "There is a wonderful sympathy and freemasonry among horsy men," remarks Sherlock Holmes in the 1891 story "A Scandal in Bohemia," and Conan Doyle detected it in fact as well as spirit among the ship captains. Peterhead was home to Keith Lodge 56 of Freemasonry's Royal Ark Mariners, established in 1739, and Manson's *Lerwick During the Past Half Century* makes clear the strong influence of Freemasonry there. Conan Doyle would become a Mason himself in 1887, though not a lifelong one.

34. The Dutch name for Novaya Zemyla, two mountainous Arctic islands that are an extension of Russia's Ural Mountains.

35. 1879 was a year of repeated tragedy for the *Ravenscraig*, a whaler out of Dundee. In May, on its way to the Davis Strait during a gale, it started taking water at an alarming rate, and some of the crew insisted upon turning back. On May 13th *The Scotsman* reported that as the ship neared the mouth of the Tay River, "the captain, after giving orders to take in sail previous to taking a pilot on board, suddenly swung himself off the bridge into the water, giving no warning of his action beyond crying 'good bye' to the mate. Two boats were lowered and manned as soon as possible, and after a quarter of an hour's search the body was recovered, but life was extinct." Later that year, under a new master, the *Ravenscraig* encountered another powerful gale, was driven onto a reef, and sank.

36. Considered more important than medical skill for ships' surgeons, because of the seal-hunting. Buchan's *Peterhead Whaling Trade*, p. 56, gives one whaler's advertisement in 1876: "A junior student not objected to, must be a good shot"

S.S. *Hope*
Lerwick
Monday
Dearest Ma'am.

here goes by the aid of a quill pen and a pot of ink to let you know all the news
from the North: the mail steamer came in yesterday with your letter and a very
kind one from that dear girl Letty, who seems to have a vague idea that I am going
to Greenland to pass an examination or face some medical board, judging from her
wishes for my success and talk about coming back quite a finished doctor. What a
jolly little soul she is though! The Scotsman came too as also did the forceps. Now
as to your inquiries I'll answer them as best I can

1st I got your letters, parcels, etc.

2nd I have not got my [?] but want it.

3rd I was not sick.

4th I have answered Mrs Hoare's letter[37]

5th I went and saw the Rodgers like a good little boy as I am. And the baby
too, at least I saw a pair of enormous watery eyes staring at me from a bundle of
clothes, a sort of female octopus with four tentacles (Octopus Dumpling = iformis).
It was far from dumb though 'Son et oculi et prosterea nihil,'[38] except a slightly
mawkish odour. Oh yes Beelzebub is a fine child – I beg its pardon – Christabel.[39]
And now that I have satisfied your perturbed spirit by soothing answers, let me
fish about for something to interest you. And first of all you will be glad to hear
that I never was more happy in my life. I've got a strong Bohemian element in me,
I'm afraid, and the life just seems to suit me.[40] Fine honest fellows the men are and
such a strapping lot. You've no idea how self-educated some of them are. The chief
engineer came up from the coal hole last night & engaged me upon Darwinism, in
the moonlight on deck. I overthrew him with great slaughter but then he took me
on to Colenso's objections to the Pentateuch[41] and got rather the best of me there.
The captain is a well informed man too. There are nearly 30 sail of whalers in
Lerwick Bay now. There are only 2 Peterhead ships, the *Windward* & *Hope*; there
is a lot of bad blood between the two sets, Gray and Murray being both looked
upon as aristocrats.[42] Colin McLean our 1st Mate was at the Queen's on Saturday

---

– "this in spite of the fact that the Board of Trade required whalers to carry a fully qualified medical practitioner." Most
were medical students from Aberdeen, Glasgow, and Edinburgh Universities.

37. Amy Hoare, wife of Dr. Reginald Ratcliffe Hoare of Birmingham, under whom Conan Doyle had externed as a
medical student. The Hoares became second parents to him, a friendship extending over many years.

38. Faulty Latin for "voice and eyes, and nothing else."

39. The Rodgers were friends of his mother's in Aberdeen. "Christabel 19!" he exclaimed in 1899 when trapped into
another visit: "Last time I saw her she was in a cradle."

40. Conan Doyle (despite evidence to the contrary) was convinced that his was a Bohemian nature, saying in *Memories
and Adventures*, about his first visit to his London uncles and aunts at age fifteen, "I was too Bohemian for them, and they
too conventional for me." He made this a theme in his first attempt at a novel in 1883, *The Narrative of John Smith* (eds.
Jon Lellenberg, Daniel Stashower and Rachel Foss; British Library, 2011), and described Sherlock Holmes so as well, Dr.
Watson saying that Holmes "loathed every form of society with his whole Bohemian soul."

41. John William Colenso (1814–83), a controversial Anglican theologian: to the dismay of fellow churchmen and many
laymen, he challenged the historicity of the Pentateuch, the first five books of the Old Testament.

42. David Gray (1828–96), master of the *Eclipse*, was in fact known as "Prince of Whalers," but by now Peterhead was

when "half a dozen Dundee officers began to run down the *Hope*. Colin is a great red bearded Scotchman of few words, so he got up slowly and said "I'm a *Hope* man mysel," and began to run amuck through the assembly. He floored a doctor & maimed a captain & got away in triumph. He remarked to me in the morning "It's lucky I was sober, Doctor, or there might have been a row." I wonder what Colin's idea of a row may be. Lerwick is the town of crooked streets and ugly maidens & fish. A most dismal hole, with 2 hotels & 1 billiard table. Country round is barren & ugly. No trees in the island. Went to Tait our agent for dinner on Friday, heavy swell feed, champagne & that sort of thing, but rather tiresome. By the way we carry capital champagne & every wine on board, & feed like prize pigs. I haven't known what it was to eat with an appetite for a long time, I want some more exercise, that's what I want. I box a little[43] but that is positively all. We just got in in time to avoid the full fury of that gale the other day. The captain says if we had stayed out we would have lost our boats & bulwarks, possibly our masts. The weather is better now. I fancy we will sail about Thursday.

There, my dear, that's about my sum total of news. God bless you all while I'm away. You'll hear from me in little more than 2 months. There is an act of Parliament forbidding us to kill a seal before April 2nd, that is why we are kicking about here. Love to all and kind remembrances to Greenhill Pl.[44]

Yr loving son

A C D

I've got the Captain's leave to go with a few of the biggest of the petty officers to the Queen's today to see if we can't have a row.

You must have got the wrong name from Mrs. Drummond.[45]

Lottie's letter was very clever & amusing.[46]

LETTER CONCLUDES, DIARY RESUMES

Tuesday March 9th

Went ashore with Captain before dinner. Jack Webster[47] was drunk and playing old Harry in the streets. Captain got hold of him and sent him on board the

---

falling behind Dundee in the industry, because Dundee's larger harbour could accommodate more and larger whaling vessels. "Peterhead still had eleven ships in 1871," says Lubbock's *Arctic Whalers*, p. 401, "but, with the exception of Gray's *Eclipse*, these were not to be compared with the fine auxiliary steamers of Dundee, and as regards whaling, Peterhead's return during the 'seventies and 'eighties was almost entirely due to the Gray brothers." (By 1880, according to Lubbock, p. 409, Peterhead's fleet was down to seven whalers, one of which could not go to sea that season.) Buchan's *Peterhead Whaling Trade*, p. 35, points out that the Grays had "decided to invest in a modern purpose-built steam vessel in an attempt to reverse the tide of poor results and financial losses."

43. In fact he boxed a good deal aboard the *Hope*, as later entries in his log will show, and "it stood me in good stead aboard the whaler," he wrote in *Memories and Adventures*.

44. Greenhill Place, Edinburgh, meant their close friends the Ryans. The widowed Mrs. Margaret Ryan was another second mother, and her son James a lifelong friend since their schooldays.

45. Charlotte Thwaites Drummond of Edinburgh, yet another second mother to him.

46. His favourite sister, six years younger than him: Caroline Mary Burton Doyle.

47. It does not become clear what place in the crew Jack Webster held, or why Captain Gray insisted upon his troublesome presence on the voyage. Efforts to find the complete crew list of the *Hope* in 1880 have been unsuccessful; it may have been among the huge number of such crew lists destroyed by the Public Records Office (now National Archives, in Kew) in recent years.

Hope in the pilot boat, but when he got half way he sprang over and swam ashore again. Cane and a boat's crew captured him afterwards. Had a very dull morning going from shop to shop. We will sail tomorrow if it is any way fair. Tait came on board afterwards and we had a pleasant talk. He is a sensible fellow tho' rather a bore. Looked over Scoresby.[48] Captain told me some curious things about whaling. The great distance at which they can hear a steamer and how it frightens them.[49] Oil is about £50 a ton and bone £800 or so. All bone goes to the continent. Sea Unicorns[50] are very common, so are sharks, and dolphins, but the curiosity of the place are the animaliculae which the whale eats.

Wednesday March 10th

A North wind prevented our getting off. The old *Eclipse* steamed in grandly about four o'clock being cheered by each ship as she passed.[51] Went on board and saw Captain David,[52] Alec and Crabbe. Went ashore in the evening and played Captain,[53] also had the honour of beating Crabbe at billiards. He has a great local reputation. Left my meerschaum and gloves in the smoking room.

Thursday March 11th

A big day for Leith.[54] The ships began to steer out from Lerwick Sound after breakfast. It was a pretty thing on the beautifully clear and calm day to hear the men singing across the bay to the clank clank of the anchors. Every ship as it passed out got 3 cheers from all the others. Captain and I went ashore, and the boat's crew and I went in search of that beggar Jack Webster. We found him at

---

48. William Scoresby the Younger (1789–1857), called "last of the great English captains" by Conan Doyle in "The Glamour of the Arctic." His father had been a highly successful whaling captain out of Whitby, England, and young Scoresby made his first voyage to Arctic seas at age ten. After scientific studies at Edinburgh University, he took over command of his father's ship *Resolution* upon turning twenty-one in 1810, with ten Arctic voyages already behind him. He spent fifteen years more devoted to not only whaling but pioneering exploration and scientific study of the Arctic, before becoming a clergyman after the death of his first wife. His logs and journals became essential guides to Arctic waters and lands, and include, as their modern editor C. Ian Jackson says in his introduction to the Hakluyt Society edition of a decade ago, "social anecdote, religious conviction, humour, and scientific inquiry." Scoresby Sound in Greenland is named for William Scoresby the Younger, and so is a crater on the moon.

49. The adoption of steam as an auxiliary to sail earlier in the century had been very controversial among whalers, and while now standard in the industry, was considered partly responsible for the overall decline in numbers of whales taken most seasons.

50. Narwhals.

51. The *Eclipse* was reckoned Peterhead's best whaler. "At a cost of almost £12,000," Sutherland's *Whaling Years*, p. 70, says, "strongly built of oak, ship rigged and heavily fortified for ice work. Her engines could generate more than 60 h.p., she carried eight whale boats, and a crew of fifty-five." She had been built in 1866 for the Grays, followed by her sister-ship *Hope* in 1872–73.

52. David Gray, also an important naturalist and authority on Arctic regions. The youngest Gray, Alexander (1839–1910), serving under David at this time, is likely the Alec referred to here. In 1883 he took the *Erik*, mentioned several times in Conan Doyle's log (usually spelling it *Eric*) – the largest whaler ever to sail from there according to Buchan's *Peterhead Whaling Trade* (p. 22).

53. Perhaps the tavern game known as The Captain's Mistress, similar to today's Connect Four, that Captain Cook played with scientists aboard his epic voyages. The 1887 edition of *Hoyle's Games* does not include a card game called Captain.

54. Unclear, though Leith is the port of Edinburgh, with a long history of whaling.

last and five of us carried him, cursing horribly, down the main street of Lerwick to the boat, where I had to hold him to keep him from jumping overboard. We left about one o'clock and steamed through the islands till about seven when we came to an anchorage with the *Jan Mayen*, *Erik*, and *Active* in a little voe.[55] We raced the *Jan Mayen* up from Lerwick and beat her all the way, anchored within a stone throw of the *Erik*. Talking to McLeod[56] and Captain about getting to the Pole in the evening. There is no doubt about it that everyone has been on a wrong tack.[57] The broad ocean is the way to find a way up to the Pole, not by going up a drain which gradually grows narrower, and down which the ice naturally runs, as it does in Davis' Straits.[58]

Friday March 12th

We'll have to stay here all day, I fear, for it is blowing half a gale tho' the glass is high. Nothing to do all day. The land is a succession of long low hills with peat cuttings and funny little thatched cottages here and there. Captain went over to the *Erik* in the evening. They seemed to be catching fish but we had no proper bait, so mate and I went ashore with a boat's crew to get some clams. It was nearly dark so we couldn't gather them, but we went the round of the little cottages begging. Such dismal hovels, the esquimeaux have better houses. Each has a little square hole in the ceiling to let out the smoke of a large peat fire in the middle of the room. They were all civil enough. Met one rather pretty but shy girl even in this barbarous spot.[59] Got some razor fish as bait and departed triumphant. Up to our thighs in mud coming and going. Revenue cutter boarded

55. Shetland term for an inlet. "It was normal practice to get clear of Lerwick," says the Shetland Library's Douglas Garden, "to anchor at one of the smaller, more northerly anchorages, to organise the crew into watches and prepare for the Atlantic crossing." Markham's 1874 journal, p. 17, jocularly suggests another possible reason: "As a general rule, the departure of a whaler is marked by the total incapacity of the crew to perform any duties whatever connected with the ship, in consequence of the numerous parting glasses of which they have partaken with their friends and acquaintances, and the bumpers they have drained to the success of the voyage."

56. John McLeod, forty-year-old First Engineer of the *Hope*.

57. How to reach the North Pole was one of the era's great challenges. "What bars the passage of the explorer as he ascends between Greenland and Spitzbergen," says Conan Doyle in "The Glamour of the Arctic," "is that huge floating ice-reef which scientific explorers have called 'the palæocrystic sea,' and the whalers, with more expressive Anglo-Saxon, 'the barrier.'" His remarks saw commentaries in a subsequent issue by Captain David Gray, Albert Hastings Markham, now a Rear Admiral who'd gone on to command the *Alert* in Britain's 1875–76 Arctic expedition, and his cousin Clements Markham, Hon. Secretary of the Royal Geographical Society and author of *Threshold of the Unknown Region*, 1873. All three agreed with Conan Doyle, with reservations, on how "the barrier" might be circumvented and the Pole reached. Despite many attempts, it would be some thirty years before a claim to have done so was widely accepted, U.S. Admiral Robert Peary by sledge with several companions in 1909. In May 1910, Conan Doyle spoke at a luncheon in his honour. "Writers of romance had always a certain amount of grievance against explorers," a newspaper reported him saying in mock-regret: "There had been a time when the world was full of blank spaces, and in which a man of imagination might have been able to give free scope to his fancy. But owing to the ill-directed energy of their guest and other gentlemen of similar tendencies these spaces were being rapidly filled up; and the question was where the romance writer was to turn."

58. The body of water between Greenland's west coast and Canada's Baffin Island. Conan Doyle expressed this view publicly on December 4, 1883, when speaking to the Portsmouth Literary & Scientific Society on "The Arctic Seas," attributing them in part to (reported the *Hampshire Telegraph*) "whaling captains who had spent their lives in those seas."

59. "Strange, barbarous, kindly people who knew nothing of the world," Conan Doyle said in *Memories and Adventures*. "I was led back to the ship by a wild, long-haired girl holding a torch, for the peat holes make it dangerous at night. I

us this evening and Lieutenant was only pacified by the present of a stick of baccy. I'm afraid Colin will eat all our bait. Captain rather annoyed about being kept in this hole. Glass high.

Saturday March 13th

Wind high and raining hard. *Active* and *Jan Mayen* are off already. We follow them soon. They are pulling up the anchor now and singing "Goodbye, Fare-thee-well, Goodbye Fare-thee-well." A pretty song it is too.[60] Sea was not very rough outside. Went through the islands, keeping full at the right at the extreme north of Shetland we passed some curious rocks in the sea called Ramna Stacks.[61]

[DRAWING 'Ramna Stacks']

Raining hard all day. We raced with the *Erik* and had rather the best of it. Not a bit seasick. Saw Burrafiord Holms the extreme north point of Great Britain,[62] and then lost sight of land about four P.M. Ran with an oblique wind and three quarter steam all night. Dreamed of being beaten by a gorilla, and of pulling in the Oxford boat.[63] 167 miles.

Sunday March 14th

Erik rather ahead of us and only occasionally in sight. Heavy Atlantic swell doing the Grand Northward Ho![64] all day under steam and sail.

[DRAWING 'Heavy Atlantic Swell']

Northward Ho! ran about 150 miles. About getting to the Pole, the Gulf Stream runs up past Spitzbergen[65] so of course that is the way to go. It is one of the most extraordinary delusions in history how ship after ship has run up into a cul-de-sac, for Davis' Straits is nothing better. Read Boswell. Don't agree with Macauley[66] at all about Boswell being a man of no intellect. If ever a man was

can see her now, her tangled black hair, her bare legs, madder-stained petticoat, and wild features under the glare of the torch."

60. A shanty about sailors heading for Liverpool and its prostitutes, the opening verse going: "Oh, we're homeward bound for Liverpool town, / Goodbye fare thee well, goodbye fare thee well, / Those Liverpool Judies they all will come down, / Hurrah, me boys, we're homeward bound!"

61. A group of skerries north of the Shetland mainland, today a protected birdlife site.

62. Another group of Shetland skerries, home to today's Muckle Flugga Lighthouse, then known as North Unst light.

63. Presumably the Oxford scull in the annual Oxford-Cambridge Boat Race going back to 1829. Conan Doyle took part in many sports throughout his life, but not to our knowledge crew.

64. Possibly referencing another book by the indefatigable Albert Hastings Markham about the Arctic, published the previous year, *Northward Ho!* (London: Macmillan), but also a popular song of 1875 about British exploration of the Arctic, "Northward Ho! or Baffled Not Beaten," lyrics by Cdr. John P. Cheyne, RN, music by Odoardi Barri.

65. The largest island of Norway's Svalbard archipelago in the Arctic.

66. Despite his criticism of Macauley here, in his 1907 book about literature and writers, *Through the Magic Door*, Conan Doyle rhapsodized over the copy he had brought on this voyage: "If I had to choose the one book out of all that line from

afflicted with what he calls "morbus Boswellianus" it is Lord Macauley himself in the case of Willy the Silent.[67]

Monday March 15th

First under steam and sail, and then under sail alone. Must have got about half way today. Kept in the cabin until evening. Read Boswell. Like that old boy Johnson for all his pomposity. A thorough old fellow, I fancy. He was in Plymouth, it seems, for a couple of days, and there was considerable ill-feeling between the townsmen and the men about the docks. Johnson who had nothing in the world to do with it was often heard to exclaim "I <u>hate</u> a docker." I like that sort of thing. Sky looked like ice this evening. Surface temperature fallen from 44 to 38 in one day.

Tuesday March 16th

Still under canvas, wind continues fair. I've brought the luck with me. Two bottlenose whales were playing round the ship in the morning but I did not see them. It seems we are crossing a very favourite feeding ground of theirs. Expect to come on the ice tomorrow. We made 159 miles yesterday. Are hundreds of miles north of Iceland, about sixty southeast of Jan Mayen. Old hands on board say they never knew such a good passage, however we mustn't crow until we are out of the wood. Water temperature has fallen 2° since 12 o'cl[ock] which looks like ice. White line on the sky. Everyone seems to think we will see ice before tomorrow. We can tell that we are under the lea of ice by the calm. Captain told me about some curious dreams of his, notably about the Germans and the black heifers.

Wednesday March 17th

Dies creta notanda.[68] About five o'clock I heard the second mate tell the Captain that we were among the ice. He got up but I was too lazy. Passed a Norwegian about 8 o'clock. When we rose at nine the keen fresh air told me it was freezing. I went on deck and there was the ice. It was not in a continuous sheet but the

---

which I have had most pleasure and most profit, I should point to yonder stained copy of Macauley's *Essays.* It seems entwined into my whole life as I look backwards. It was my comrade in my student days, it has been with me on the sweltering Gold Coast, and it formed part of my humble kit when I went a-whaling in the Arctic. Honest Scotch harpooners have addled their brains over it, and you may still see the grease stains where the second engineer grappled with Frederick the Great. Tattered and dirty and worn, no gilt-edged morocco-bound volume could ever take its place for me."

67. William I, Prince of Orange (1533–84), who led the Dutch revolt against Spanish rule. In his *History of England*, Macaulay took a dim view of him. His great-grandson William of Orange became King of England and Scotland in 1689 following the overthrow of the Stuarts.

68. "A joyful day for Crete," from one of Horace's Odes.

whole ocean was covered with little hillocks of it, rising and falling with the waves, pure white above and of a wonderful green below.[69] None were more than 4 or 6 feet out of the water but they were of every shape. No seals. Put up the crow's nest in the morning.

[DRAWING 'A Peterhead Whaler' SUBTITLED '(Ice in the background by Capt John Gray of the *Hope*)']
[DRAWING 'Sealing Costume']

All day we were steaming or rather sailing through lumps of ice which studded the water, sometimes so thickly that you could jump from one to another for hundreds of yards, and sometimes only a bit or two visible. The large ice field seems to be on our left. See a ship about 5 miles behind us, supposed to be the *Jan Mayen*, while far away in front a sail is dimly visible. From the masthead Cane says he can see 9 vessels.

Thursday 18th March

Stewart dreamed that he was among a great herd of swine last night, so we are sure to see seals today. If a man dreams of anything agricultural it always means that seals are somewhere near. A curious fact. Ice lying in lumps much the same as yesterday. Stewart's dream seems true for we saw our first seal, a bladdernose, about 11 AM. It was speckled black and white and lay on the ice as the ship steamed past, only about a dozen yards from it, looking at it quietly. Poor brute, if they are all as tame it seems a shame to kill them.

[DRAWING 'our first seal']

Captain saw a large speckled owl a couple of hundred yards from the ship, saw a few roaches and guillemots[70] but we are too far from land to have many. We are considerably to the North of Jan Mayen now. Passed another bladdernose and a saddleback seal later. Some were seen in the water afterwards. A most lovely morning but hazy towards evening. Spoke to the *Erik* and mutually congratulated

---

69. Addressing the Portsmouth Literary & Scientific Society in 1883, he invited his audience, reported the *Portsmouth Times*, to imagine the voyage out: "after about four days' sailing they would leave behind them the Unst light, the last trace of civilisation they would see…. another twenty-four hours' sailing brought them to the latitude of Iceland, and then the voyagers met with streams of drift ice separated from the main pack, the effect of this change on the unaccustomed traveler being graphically described. After the 76th degree … they got into the floes, great floating plains, some of them as large as an English county, and none smaller than Southsea Common. The dangers of navigating the leads, or narrow lanes of water between the floes, having been referred to, Dr. Doyle brought his hearers into the whaling grounds, where the water was as thick as pea soup with the millions of minute creatures that nourished the great cetacean. Having by dint of many disappointments got as far as between the 80th and 81st parallels they found from east to west one great wall of ice, without a break, extending from the north end of Spitzbergen to the east coast of Greenland. They might here put their yacht southwards without any humiliation," he concluded, "for they had come within 120 miles of the highest point ever reached by a European ship."

70. Roaches and guillemots: varieties of seabird.

each other on our passage. By the way Walker said to me at Lerwick "If I had known who you were, sir, last year, things might have been different." I'm a lot better as I am, though I didn't make that remark to him.[71]

Friday March 19th

A thick haze with the lumps of ice looming out of it. Could see about a hundred yards in each direction. Passed two large bladdernoses, male & female on a bit of ice.

[FOLDED-OVER DRAWING "The Hope among loose ice March 16th 1880"]

We tried the whistling and certainly the male did stop and listen to it, the female wasn't so susceptible but shunted at once. The male was about 10 feet long, I should think, the female 7 or 8. I wish the haze would clear up. Drizzling a little. Haze continued all day so we lay to at night. Cane and Stewart were sparring in the evening. Talk on literature with the Captain, he thinks Dickens very small beer beside Thackery. Buckland seems to be a lovely sort of cove.[72]

Saturday March 20th

Only a week from Shetland and here we are far into the icefields. It has certainly been a splendid voyage. Beautiful day, wonderfully clear. Icefields, snow white on very dark blue water as far as the eye can reach. We are ploughing through in grand style. Five sail in sight, one the *Erik*. Stewart insists on my accepting a pretty Esquimeaux tobacco pouch; I suppose he means it as a quid pro quo for the pipe I gave him. No seals seen as yet. Got near heavy ice in the evening and lay to. Several bladdernoses playing about the ship. About a couple of hundred seals visible from the crow's nest, so we seem to be coming near the pack. Eleven sail in sight. Adam Carner saw the steps of a bear in the ice.[73]

Sunday March 21st

Lay to all day owing to the thick haze. Bladdernoses by the dozen are all around us. A few saddlebacks. The Captain thinks the pack is about 20 miles or so in

---

71. Dr. Robert Walker of the *Eclipse* may have been, in shore-based practice, one of the doctors to whom Conan Doyle had applied in 1879, unsuccessfully, for an externship as a medical student.

72. Francis Trevelyan Buckland (1826–80), English physician and naturalist. "Four and a half feet in height and rather more in breadth," said an acquaintance, with a bouncy writing style in books like *Log Book of a Fisherman and Zoologist* (1876) that made him a popular author and lecturer in the 1860s and '70s. He was an advocate of zoöphagy, i.e. the eating of exotic animals, reptiles, birds, etc., and kept a remarkable menagerie at his London home to supply his kitchen. As Her Majesty's Inspector of Fisheries during the 1870s, he came to know the Grays, especially Captain David; the subsequent ban on sealing before April 3rd each year was largely their work, as a conservation measure, according to Gavin Sutherland's *Whaling Years*, pp. 91–92. Buckland praised Gray in his posthumously published *Notes and Jottings from Animal Life* (1882).

73. While Conan Doyle gives the name as Carner throughout the log, we believe this to be Adam Cardno, a Peterhead native born 1833, and a boatswain who reached the rank of Fast Harpooner no later than 1871, when serving under Captain Gray on an earlier vessel.

front of us. Johnny[74] had a meeting in the evening, the singing sounded well from the deck.[75] Split a bottle of port after dinner. Captain tells me he tried fixing a cone full of prussic acid onto the end of the harpoon. He fired it into a finner from his small steamer. The brute went away at such a rate that it very nearly set the bows on fire by the friction. The line broke and it got away, but seems to have died, for no dogfish were seen on the coast for some days. Many finners are 100 feet long. By the way Carner taught me some esquimeaux. Amalang (yes), piou (very good), piou smali (bad), kisi-micky (ice-dog – ie bear).

Monday March 22nd

Very foggy again, but we have drifted among a few saddlebacks with their little fat yellow offspring. Got the quarter boats out, and the rifles. A long time to wait yet, though, till April 3d, Saturday week. Fog lasted all day so that we lay to. Boxed in evening. Finished Boswell Vol I. Dreamed of G. P.[76]

Tuesday March 23d

Clear morning, a good few seals in sight.

[DRAWING 'male, female & young saddleback']

Eclipse came in at last, and Captain boarded it before dinner. Steamed a few miles in the right direction. Blowing a gale all day. 11 degrees below freezing point. Very cold wind. Rigging covered with ice. Climbed up to the crow's nest before tea, but the Captain called me down just as I got up to it, as he thought I might get frostbitten. Got a fine pouch from Cane. Carner tells me at New Orleans before the war a dock labourer could make £1 a day. Now they make a dollar only. Captain saw blockade runners[77] leaving Liverpool during the war, long spider like steamers of great speed, and painted the colour of the ocean. Cargo mostly quinine, needed hardly any crew. Glass rising again.

---

74. The chief engineer, John McLeod.

75. Conan Doyle has not previously mentioned church services, but Sutherland's *Whaling Years*, p. 27, says: "The 'whale boys' were a rough-and-ready bunch, deeply superstitious, and God fearing to a man. A Sunday never passed without a time of worship, conducted in accordance with the Church of England's 'Seaman's Prayer Book.' Though very few of the crewmen were of that denomination, the custom was adopted in order that no favour be shown to either the Presbyterian or Roman Catholic company. With some ships having Divine Service both morning and afternoon, and Bible readings in the evenings, it was not unknown for the meetings to be abruptly interrupted by the 'blow' of a whale. Men would rush to the boats Bible in hand and, after slaughtering the poor beast, return to psalms and prayer."

76. Unclear; perhaps the initials of a girl on whom he was sweet, or perhaps only looking forward to his time as a qualified physician in general practice.

77. Confederate vessels of the American Civil War, 1861–65: built for speed to evade the Union blockade of Southern ports, and bring war-critical commodities like quinine to the Confederacy.

Wednesday March 24th

Another big day for Leith. We have seen the pack, and an enormous pack it is too. I have not seen it from the nest yet but it extends from one side of the horizon to the other, and so deep that we can see no end to it. The nearer we steam towards it, the bigger it grows. Colin says he never saw such a one in his life. It is certainly the largest collection of big animals in the world at present, at least I know no other beast that goes in herds of millions, covering a space about 15 miles long and 8 deep. We ought to have a good voyage now, my old luck. All the ships are lying round now and taking up their positions. *Windward* steamed past us today flying her Jack, and dipped it as a salute. 10 days yet to wait. Oysters.

Thursday March 25th

Hurrah for a quill pen! 19° below freezing point this evening. Have been taking up our position, and mounting boats and cleaning guns all day. Edge of pack can be seen from the bridge now. Good many isolated ones about the ship. I can hear the young ones squeaking as I write. It is a noise between the mew of a cat and the bleat of a lamb. They look a sort of cross between a lamb & a gigantic slug. Our only fear now is that some of these great blundering Norwegians or Dundeesmen go and put their foot into it. If we get less than 50 tons I'll be disappointed, if we get less than 100 I'll be surprised. Captain is going to teach me to take the latitude and longitude. Saw a clever couplet today

> "Till Silence, like a poultice comes,
> To heal the blows of Sound."

Holmes' I think.[78] Sported my sea boots today.

Friday March 26th

Frost still continues, 17° today, 20° during the night. This is just what we want to fill up gaps in the icefield and make it safe walking. Steamed very little. The mate says the seals are lying in an almost solid mass. He says there are more than in '55,[79] and in that year 50 vessels were among them, and all got filled. We are 23 vessels now all told so the prospect is cheery. Bar earthquakes we'll make a voyage of it. It is very trying work waiting, though this close time is an excellent

---

78. Oliver Wendell Holmes of Boston, a favourite writer of his, recalling here lines from "The Music-Gardens" (1836): "And Silence, like a poultice, comes / to heal the blows of sound."

79. 1855 was the peak year for Scotland's seal industry, with 131,049 seals taken by twenty-seven Peterhead vessels, according to *The Scotsman* ("The Peterhead Whalers," November 19, 1902).

provision. The poor brutes used to be killed before they had pupped. *Eclipse* got a bear today, and we saw the steps of one on the snow beside our ship. They are cowardly brutes unless in a corner. Captain killed one once with a boat hook. Engineer told me how one chased a crew for miles across the snow once, and how they had to throw down article after article to engage his attention, so that they got to the ship nearly naked and in a blue funk. There is no specimen of a right whale in any British museum, except a foetus. Saw the young seals suckling today. Hurt my hand boxing with the Stewart. Stuffed old Keith's tooth, and cured young Keith's collywobbles.[80] It seems to be the family's day out.

Saturday March 27th

This day week is our day. Got my knife and my sharpener today, and asked Carner to see about my club.

[DRAWING 'Knife']

Beautiful day, still lying on the skirts of the pack, all seem satisfied except the Captain and he grumbles a bit, but I think he is only joking. Saw another bear's footsteps.

[DRAWING 'Bear's Step']

The *Eclipse* has killed two and we have never seen one. They tell me bears go in flocks of 20 or 30 very often. Rifles given out tonight. Steamed a little. Haggie Milne better tonight. No news. Wrote my "Modern Parable."[81]

Sunday March 28th

Haggie bad again so I gave him some Chlorodyne.[82] Captain went on board the *Eclipse* and in a little the boat came off for me for dinner. Had a very pleasant feed with good wine afterwards.[83] The conversation turned upon the war,[84] politics, the North Pole, Darwinism, Frankenstein,[85] free trade, whaling and

---

80. A 19th-century colloquialism for pain or looseness in the bowels.

81. Lost. Conan Doyle had published a story the year before, in *Chambers' Journal*. "The Mystery of Sasassa Valley" was slight, and his next efforts were rejected, but "it mattered not that other attempts failed. I had done it once and I cheered myself by the thought that I could do it again."

82. Medicine that sounds like cure-or-kill: "A popular anodyne composed of chloroform, morphia, Indian hemp, prussic acid, etc.," says the Oxford Universal Dictionary.

83. "Even at the highest polar latitudes, at Captain's table," says Sutherland's *Whaling Years*, p. 27, "the strictest rules of Victorian etiquette were religiously observed as guests enjoyed freshly shot sea-fowl, usually eider duck or diver, washed down with a fine claret, and followed by brandy, conversation, and perhaps a rubber of whist."

84. The Second Afghan War, 1878–80. "We had left in exciting times," Conan Doyle remembered in *Memories and Adventures*. "The Afghan campaign had been undertaken, and war seemed imminent with Russia. We returned opposite the mouth of the Baltic [in August] without any means of knowing whether some cruiser might not treat us as we had treated the whales."

85. Mary Shelley's novel *Frankenstein* (1818) begins and concludes aboard an Arctic exploration vessel, and is narrated by its captain who previously had served on Greenland whalers. Shelley, prior to her marriage to the poet, had spent

local matters. Captain David seems to take a sinister view of our case. Says we'll be lucky if we get 20 tons; he may say it, but I don't think he thinks it. Saw his bear's skins. By the way he told us some strange stories which I will try to write as he told them.

"When I was a young fellow," he said "I happened to be in London with a gold watch and a good deal of money. I was at the Lyceum[86] one night and wanted to get back to my lodgings in Holborn[87] but wandered about a long time unable to find my way. At last I saw a respectable looking man and asked him the way to Holborn, adding that I was a stranger. He said he was going that way himself, and that he was Captain Burton of the 17th Lancers.[88] We walked on together and Captain Burton by turning the conversation on the danger of carrying money about in London, learned about my watch and gold, and warned me against it. We shortly afterwards turned into an open door and the Captain said 'What shall we have here, I'll have some Cognac.' I said 'Coffee is strong enough for me.' The waiter who brought in the things was the most repulsive looking ruffian I ever clapt eyes on, and I saw him stick his tongue in his cheek and leer at the Captain. It was then that I first suspected that I had got into a trap.

"I threw half a sovereign on the counter and rose to go out, but the waiter put his back against the door and said 'We don't allow our visitors to leave us like this.' The Captain said 'Come on, sir, and we'll make a night of it; hullo give us some sherry out of bin No 3.' The waiter called 'Janet' and a girl appeared rather pretty and very pale. He said 'Bin No 3.' The girl said 'Surely, surely you don't need that bin tonight.' He said 'do what you are told.' As she brought in the wine she whispered to me 'Pretend to sleep.' I drank a little of the wine, but spilled most of it. Then I sank down & closed my eyes. Soon the two villains came over and whispered together, and one passed the candle over my eyes and said 'He is off.' They whispered a little again, and one said 'Dead men tell no tales.' The other said 'Then we had better get the bed ready' and they both left the room. I flung open the window and was off down the street like a shot, and ran about half a mile before I saw a bobby, and then I found it impossible, with

substantial periods of 1813 and 1814 in Dundee, and heard much about its whaling industry. She used what she learned to describe in ch. 24 of her novel the vessel's being beset by the Arctic ice that threatens to crush its hull and destroy all aboard, including the dying Dr. Victor Frankenstein, whose monster lurks nearby.

86. One of Victorian London's chief theatres, Wellington Street, Strand. In *The Sign of Four*, the second Sherlock Holmes tale (1889), Holmes and Dr. Watson rendezvous there with Miss Mary Morstan, to meet her mysterious benefactor and begin the adventure.

87. The extension of Oxford Street from London's West End to the City. Its historical and literary significance made it useful to two of Conan Doyle's Sherlock Holmes stories.

88. The Duke of Cambridge's Own, whose motto "Death or Glory" had been realised with heavy losses in 1854 during the Crimean War as part of the Charge of the Light Brigade.

my imperfect knowledge of London to find the house again. I heard no more of it. Get out another bottle of Port, Doctor." The conclusion of the story was considered to be a very able effort.[89] He told us another story about how he acted as a spy in the Boer service,[90] and murdered 3 Kaffirs[91] in their sleep, and shot a German through the body.

He saw a walrus eating a Narwhal once. He is a fine fellow, and Dr Walker seems a very decent chap too. He thinks more whales are found at night than in the day, so when he gets North into the Twilight land, he has his breakfast at 10 P.M., dinner at 2 in the morning, and supper at 7 A.M. Then he sleeps all day. He says whales leave a very characteristic odour behind them, and you often smell them before you see them.

March 29th Monday

Our time is coming now. Thick day with a driving snow. Nothing particular going on. Had a pleasant evening in the mates' berth. Songs all round. Sang "Jack's Yarn," "The Mermaid" and "Steam Arm."[92] Good fun. By the way Colin the mate paid me a high compliment today. He said "I'm going to have every man working hard when we start sealing. I've no fears of you, Surgeon. I'll back you to do a day's work with any man aboard. You suit me, and I liked the style of you the first time I saw ye. I hate your clean-handed gentlemen." This was a high compliment from taciturn Colin.

March 30th Tuesday

Nothing much doing. *Windward* came alongside and Murray came on board. He seemed to have small prospects, 10 tons was more than he expected, he said. Told us about Sir John Ross firing his gun through the window of a house because his mate was inside & he wanted him. Murray was one of the Franklin searchers. Ross said "Every step onwards, boys, is honour and glory to us. Death before dishonour," when they were starting sledging.[93] Sparred with Colin & Stewart.

---

89. The account is reminiscent of Robert Louis Stevenson's *Suicide Club* trilogy, 1878, collected with other tales as *New Arabian Nights*, and may have struck Conan Doyle because he did not yet know London well, but was intrigued by this side of it. Visiting relatives in 1874, at age fifteen, he had written home in delight about the Chamber of Horrors at Mme. Tussaud's wax museum in Baker Street; and Captain David's tale also likely echoed when six years later he created Sherlock Holmes whose adventures often took him into the "lowest and vilest alleys in London."

90. Boers: the Dutch settlers of South Africa whose Transvaal Republic and Orange Free State rivaled Britain's Cape Colony. The First Boer War broke out that December, ending indecisively in March 1881. In late 1899 the Second, or Great Boer War broke out, lasting several bitter years, with Conan Doyle going to South Africa as a volunteer army field surgeon for six months.

91. A derogatory term for South African native peoples.

92. There have been too many sailor songs about mermaids to say which one this may have been, but "Steam Arm," an early-19th-century song by H. V. Smith, was about a soldier returning from Waterloo with only one arm, and building a mechanical steam-powered one in order to cope with his shrewish wife.

93. Admiral Sir John Ross (1777–1856), Scottish explorer whose expeditions to find a Northwest Passage were unsuccessful

Wednesday March 31st

Very little doing all day. A heavy swell has set in and we are uneasy about the result. If it continues until Saturday it will make our work both difficult and dangerous. The ice is not a solid sheet, but made up of thousands of pieces of all sizes floating close to each other. Now in a swell those pieces alternately separate and come together with irresistible force. If a poor fellow slips in between two pieces as is easily done, he runs a good chance of being cut in two, as actually happened to several Dundeesmen.[94] Men played leapfrog on a big piece. I started a story "A Journey to the Pole," which I intend to be good. We are going to write to Gladstone and Disraeli when the Dundeesmen go home.[95]

Thursday April 1st

Swell continues and things look badly. We steamed a bit during the day. This is the first time for 3 years that I have not been examined today.[96] Sent the Chief Engineer to the Captain with a cock and bull story about curtain rings. Johnny's dignity was very much hurt. By the way I was at the masthead yesterday, and also on the ice some time. Saluted the *Harald Haarfager*[97] tonight 7.30. Swell still on.

[FOLDED-OVER DRAWING "Ships taking up their positions among the seals"]

Friday April 2nd

Swell still on and the pack growing more scattered. I'm afraid our prospects will not be realized. However every man must do his best, and then we can do no more. Stayed up until 12 o'clock to see the close time out.

Saturday April 3rd

Up at 2.30 AM. Swell still on, so as to make good work impossible. Lowered away our boats in the sludge about 4.30. I stayed aboard at the captain's command much against my will and helped as well as I could by pulling the skins up the

---

but produced much valuable scientific information. In 1850 he led an attempt to rescue Admiral Sir John Franklin's 1845 expedition to Canada's Arctic waters, but the party had perished. Ross had once been stranded in the Arctic with his crew for four years.

94. J. F. Taplin of the *Windward* bore witness to such fatalities in his diary (April 15, 1860): "The poor fellow had fallen through a piece of ice and whilst he was trying to save himself a gust of wind came and sent another piece of ice against him, this about cut the poor fellow in half ."

95. Benjamin Disraeli, a Conservative, had been Prime Minister since early 1874, but in less than a month would be succeeded by the Liberals' William Ewart Gladstone. Perhaps the idea was to ask these statesmen to have the Dundee whalers suppressed.

96. Referring to medical school examinations at Edinburgh University.

97. A Norwegian ship named for Harald Fair Hair, the Viking chieftain who united the Norwegian lands and became king of Norway in the year 872.

side.[98] The old seals who can swim are shot with rifles, while the poor youngsters who can't get away have their skulls smashed in by clubs.

[DRAWING 'Seal Club']

It is bloody work dashing out the poor little beggars brains while they look up with their big dark eyes into your face. We picked the boats up soon and started packing, that's to say all hands getting over the ship's side and jumping along from floating piece to piece, killing all they can see, while the ship steams after and picks up the skins. It takes a lot of knack to know what ice will bear you, and what not. I was ambitious to start but in getting over the ship's side I fell in between two pieces of ice and was hauled out by a boathook.[99] I changed my clothes and started again, & succeeded in killing a couple of seals and dragging their remains after I had skinned them to the ship's side. We got 760 seals today. Poor work, I believe but we hope for the best. After all whales are the things that pay.

Sunday April 4th

Working all day. I fell into the Arctic Ocean three times today, but luckily someone was always near to pull me out. The danger in falling in is that with a heavy swell on as there is now, you may be cut in two pretty well by two pieces of ice coming together and nipping you. I got several drags, but was laid up in the evening as all my clothes were in the engine room drying. By the way as an instance of abstraction of mind, after skinning a seal today I walked away with the two hind flippers in my hand, leaving my mittens on the ice. Some of our hands work very well, while others, mostly Shetlanders with many honourable exceptions, shirk their work detestably.[100] It shows what a man is made of, this work, as we are often killing far from the ship away from the Captain's eye with a couple of miles drag, and a man can skulk if he will. Colin the mate is a great power in the land, energetic & hard working. I heard him tell a man today he would club him if he didn't work harder. I saw the beggars often walk past a fine

98. As noted in this volume's introduction, Conan Doyle was still smarting about it when he wrote "Life on a Greenland Whaler" years later, but also admitted there that: "I justified his original caution by falling in twice again during the day, and I finished it ignominiously by having to take to my bed while all my clothes were drying in the engine-room."

99. No trifling matter, in sea-water near freezing. "Look here," says a character in Conan Doyle's 1882 story "The Captain of the Pole-Star," "it's a dangerous place this, even at its best – a treacherous, dangerous place. I have known men cut off very suddenly in a land like this. A slip would do it sometimes – a single slip, and down you go through a crack, and only a bubble on the green water to show where it was that you sank." Two days after this entry, Conan Doyle came close to death this way.

100. Conan Doyle took a more equitable view afterwards. In "Life on a Greenland Whaler" sixteen years later he said: "There were fifty men upon our whaler, of whom half were Scotchmen and half Shetlanders whom we picked up at Lerwick as we passed. The Shetlanders were the steadier and more tractable, quiet, decent, and soft-spoken; while the Scotch seamen were more likely to give trouble, but also more virile and of stronger character. The officers and harpooners were all Scotch, but as ordinary seamen, and especially as boatmen, the Shetlanders were as good as could be wished." The full crew was fifty-five or –six hands of which nineteen were Shetlanders.

fat seal to kill a poor little "Toby" or newly pupped one in order to have less weight to drag. The Captain sits at the masthead all day, looking out with his glass, for where they lie thickest. Took about 460 today.

Monday April 6th [5th]

Went out with Colin this morning for some regular hard work but began proceedings by falling into the sea again. I had just killed a seal on a large piece when I fell over the side. Nobody was near and the water was deadly cold. I had hold of the edge of the ice to prevent my sinking, but it was too smooth and slippery to climb up by, but at last I got hold of the seal's hind flippers and managed to pull myself up by them. The poor old "flappy" certainly heaped coals of fire upon my head.[101] Got off again with the Stewart and did some good work.[102] Took about 400 again.

Tuesday April 6th

Out on the pack in the morning with Colin and actually did not fall in. The Captain calls me "the Great Northern Diver." We took a good number of young and old and then steamed outside to see if we could find anything for ourselves. Shot two large bladdernoses, both were easy shots at about 70 yards, but as I fired after all the harpooners had missed I felt cocky. They were huge brutes, I am keeping the bone of one which was 11 feet long. They are also called Sea Elephants. They have a vascular bag on their snouts which they distend to any extent when they are angry. Saw the *Jan Mayen* and others, with all their boats out killing old seals. Took 270 young & 58 old.

Wednesday April 7th

Poor work today, seals are scarce and we only took 133. Haggie Milne is very bad & I fear he will die. He has intussusception[103] with foecal vomiting & constant pain. It is not hernia. Gave soap & castor oil injection today.[104]

---

101. Dr. Taplin in 1860 had had a similar mishap, and, though pulled out by a nearby mate, found getting back to his ship no joke: "I had to run very fast to prevent my legs from becoming frozen, the ice was very bad but I arrived at the ship in about 40 minutes. I came at once down into the cabin to change but it was some time before this could be done as my pantaloons were frozen to my boots and my coat was frozen to both my waistcoat and pantaloons. I cannot think how I got to the ship … when I reached her my legs were so stiff I could not get to the deck by the ladder but had a rope placed round under my arms and got hoisted to the deck."

102. Conan Doyle was young, vigorous, and fortunate. Albert Hastings Markham speaks, p. 77, of an Arctic seaman who fell into the icy sea similarly: though "taken on board and restoratives administered … it was many days before he recovered from the effects of his cold bath."

103. An intestinal disorder in which part of the intestine slides into another part of it, preventing food or fluid from passing through and cutting off blood to the affected part. Fatal if not treated promptly, with surgery required as the ultimate resort, but beyond the capacity of an 1880 third-year medical student aboard a whaling ship with few facilities and no assistance.

104. According to Dr. Robert Katz, "probably given as an enema, one treatment for that diagnosis. However, intus-

Thursday April 8th

Put our letters on board the *Active* today.[105] Had short notice and only wrote one letter though I would willingly have written more. Did a wretched days work, only about 30 seals. However most of the other ships have done worse than us, & that with crews of 80 men to our 56. Gale in the evening.

LETTER TO HIS MOTHER IN EDINBURGH

Latitude 73° 10 N. Longt. 2° E.
S.S. *Hope.*
Greenland. April 7, 1880

Dearest Ma'am

Here I am as well and as strong and as ugly as ever off Jan Mayen's Island in the Arctic Circle. We started from Shetland on the 10th of March, & had a splendid passage without a cloud in the sky, reaching the ice upon the 16th. We went to bed with a great stretch of blue water before us as far as the eye could reach, & when we got on deck in the morning there was the whole sea full of great flat lumps of ice, white above & bluish green below all tossing and heaving on the waves. We pushed through it for a day but saw no seals, but on the second day we saw a young sea elephant upon the ice, & some schools of seals in the water swimming towards N.W. We followed their track & on the 18th saw the smoke of 6 steamers all making in the same direction, in the hope of reaching the main pack. Next morning eleven vessels could be seen from the deck, & a lot of sea elephants or bladdernose seals so we felt hopeful. You must know that no blood is allowed to be shed in the Arctic Circle before April 3rd.[106] On the 20th March we saw the real pack. They were lying in a solid mass upon the ice, about 15 miles by 8, literally millions of them. On the 22nd we got upon the edge of them and waited. 25 vessels were in sight doing the same thing. On the 29th a gale broke and the pack were sadly scattered, and a couple of Norwegian lubbers came steaming through them, frightening those that had not pupped away. On the 3rd the bloody work began and it has been going on ever since. The mothers are shot & the little ones have their brains knocked out with spiked clubs. They are then skinned where they lie & the skin with blubber attached is dragged by the assassin to the ship's side. This is very hard work, as you often have to travel a couple of miles, as I did today, jumping from piece to piece before you find your victim, & then you have a fearful weight to drag back. The crew must think me a man of extraordinary tastes to work hard & with

susception is a difficult diagnosis to make based on physical examination alone. Even with X-rays, the diagnosis is often confirmed only at surgical exploration. In addition, this is a disorder more common in children, and the patient was an older man. [Milne was seventy years old.] It sounds as if Milne had some type of intestinal obstruction, which has a rather lengthy differential, including adhesions, vascular insufficiency, or tumor (all more likely in an adult). In the modern era, the patient would be fed intravenously until the condition was treated."

105. According to Buchan's *Peterhead Whaling Trade*, p. 54, John Gray's first command – when but twenty-two years old.

106. By agreement since 1877 between Britain and Norway, the conservation measure brought about on the British side by Captain David Gray and Fisheries Inspector Frank Buckland.

gusto at what they all consider the most tiring task they have, but I think it encourages them. My shoulders are all chafed with the Lourie-tow or dragging rope.[107] By the way in the last four days I have fallen into the sea five times which is a pretty good average. The first time I tried to get onto the ice, there was a fine strong piece alongside, and I was swinging myself down onto it by a rope, when the ship gave a turn of her propeller sending me clear of the ice and into the sea with 28° of frost on. I was hauled out by a boathook in my coat, and went on the ice again when I had changed, without mishap. I was not so fortunate next day for I fell in three times & all the clothes I had in the world were in the engine room drying. Next day I fell in once, and now I have had two days of immunity. It takes considerable practise to know what ice is trustworthy & what is not. We have seen the steps of bears in the snow about the ship but I have not had a bang at one yet. I shot a fine sea elephant yesterday 11 feet long, as big as a walrus. They are formidable brutes and can give a bear more than he brings. Our young sealing is over now & has been a comparative failure, about 25 tons, but we will follow up with the old seals now as they go North, and then away we go past Spitzbergen & over 80° Lat for the whaling where we hope to do better. I have enjoyed my voyage immensely, my dear, and only hope you are as cheery. I don't think you would have recognized me as I came into the cabin just now – I'm sure you wouldn't. The captain says I make the most awful looking savage he ever saw. My hair was on end, my face covered with dirt and perspiration, and my hands with blood. I had my oldest clothes on, my sea boots were shining with water and crusted with snow at the top. I had a belt round my coat with a knife in a sheath and a steel stuck in it, all clotted with blood. I had a coil of rope slung round my shoulders, & a long gory poleaxe in my hand. That's the photograph of your little cherub, madam. I never before knew what it was to be thoroughly healthy. I just feel as if I could go anywhere or do anything. I'm sure I could go anywhere and eat anything. Now, my dear, don't be uneasy during the next month or two. If ever a round peg (not pig) got into a round hole it is me. Give my love to Greenhill Place, Mrs Waller & the Doctor,[108] also to Mrs Neilson & all in London.[109] I would have written to Greenhill Pl and London but there is a ship alongside for our letters & I thought one good is worth three bad ones.

    All kind regards to Mrs Budd and Budd himself.[110] Don't lose his address.

---

107. Conan Doyle understates the difficulties. In his 1883 talk reported by the *Portsmouth Times*, he told the audience: "If you could conceive the space between Portsmouth and London to be divided every three or four paces into deep rugged chasms and slits too broad to jump across; if you then sprinkle promiscuously among and on top of these chasms millions of great boulders averaging the size of a house; in some places let the sloping surfaces be as slippery as glass; in others let there be snow in which one sinks to one's neck; add to this a little Arctic climate, in which the least exposure is followed by frostbite; and then start on this journey hauling a dead weight of 100 or 150 pounds behind you, sliding it down one side of a chasm and hauling it up the other, climbing up the great boulders and dragging it up behind you: before you reached London you would have a pretty clear conception that distance in the Arctic seas is not to be computed by miles, and that less labour would take a man from London to Constantinople than is represented by those sixty miles which separate us from the hardiest of our foreign rivals."

108. Dr. Bryan Charles Waller, six years older than Conan Doyle, and an important, sometimes turbulent influence on the latter's education and medical career; perhaps a literary influence as well, as Waller was a published poet. While there were ruptures from time to time, Waller was Best Man at Conan Doyle's wedding in 1885.

109. "All in London" would have included uncles and aunts there, but Mrs. Neilson is unknown.

110. Dr. George Turnavine Budd, a year ahead in medical school. Despite a volcanic personality and controversial ways, Conan Doyle agreed, against advice from others, to become Budd's assistant in Plymouth in 1882. "A brilliant

Yr loving son

Arthur C. Doyle

The Captain sends his compliments & says that I am an untidy rag; but sternly refuses to explain the meaning of this term of opprobrium. He calls me the "Great Northern Diver" too in allusion to my recent exploits in the bathing way. I have never had a touch of sea sickness since I left Edinburgh. Love to Papa, Mary Lottie & Connie.[111]

LETTER CONCLUDES, DIARY RESUMES

Friday April 9th

Gale continuing so that we have done no work at all. Heavy swell on. Got under the lee of the point. Wretched day. Did nothing but sleep & write up my log. They are commencing to cut the blubber off the hides. I'm afraid tomorrow will be as bad.

[EIGHT SMALL DRAWINGS, OF 'Dragging Seal Skins,' 'Waiting for the mother,' 'a procession,' 'Clubbing a young one,' 'a big load,' 'my accident,' 'a dangerous bit,' AND 'Flinching a seal']

April 10th Saturday

Poor Andrew Milne is almost beyond hope. At such an age and with such an illness recovery was almost hopeless. Blowing fitfully and with a heavy swell on. Nothing doing all day. Began Carlyle's "Hero Worship."[112] A great and glorious book.

April 11th Sunday

A dark day in the ship's cruise. Poor Andrew was very cheery and very much better in the morning, but he took some plum duff at dinner, and was taken

---

career seemed to lie before him, for besides his deep knowledge of medicine, acquired in the most practical school in the world, he had that indescribable manner which gains a patient's confidence at once," was how Conan Doyle saw it in a semi-autobiographical short story, "Crabbe's Practice." In Plymouth he found Budd had established a bustling practice, but by methods outside professional ethics. The association lasted six weeks. Conan Doyle moved on to Southsea, Portsmouth, to start a practice of his own. In a later novel, *The Stark Munro Letters*, and in *Memories and Adventures*, he wrote about the uproarious association (calling Budd Cullingworth). Budd died at age thirty-four, apparently from a brain abnormality that may have influenced his behaviour.

111. His sister Lottie, mentioned before; Connie was another, age twelve at this time. Mary, however, was a sixteen-year-old Irish domestic servant in the Doyle household, Mary Kilpatrick.

112. Scottish historian Thomas Carlyle (1795–1881) was an exceptionally important writer for the young Conan Doyle, if not without reservations – saying in a diary during another sea voyage the following year that Carlyle was "a grand rugged intellect [but] I fancy Poetry, Art and all the little amenities of life were dead letters to him." Carlyle's *On Heroes, Hero-Worship, and the Heroic in History*, 1841, covered men as diverse as Cromwell, Shakespeare, Napoleon, Johnson, Martin Luther, and Mohammed. Conan Doyle probably liked Carlyle's remarks that "All that mankind has done, thought, gained or been: it is lying as in magic preservation in the pages of books" and "What we become depends on what we read after all of the professors have finished with us. The greatest university of all is a collection of books" – much his own view in his book *Through the Magic Door* in 1907. Carlyle was neglected in it, but played a significant part in Conan Doyle's first attempt at a novel in 1883 (*The Narrative of John Smith*).

worse. I went down at once, and he died within ten minutes in my arms literally.[113] Poor old man. They were very kind to him forwards during his illness, and certainly I did my best for him.[114] Made a list of his effects in the evening. Rather a picturesque scene with the corpse and the lanterns and the wild faces around. We bury him tomorrow. Picked up seals all day on large pieces in the slush, about 60 I think.

April 12th Monday

Buried poor old Andrew this morning. Union Jack was hoisted half mast high. He was tied up in canvas sack with a bag of old iron tied to his feet, and the Church of England burial service was read over him. Then the stretcher on which he was lying was tilted over and the old man went down feet foremost with hardly a splash. There was a bubble or two and a gurgle and that was the end of old Andrew. He knows the great secret now. I should think he would be flattened out of all semblance to humanity before he reached the bottom, or rather he would never reach the bottom, but hang suspended half way down like Mahomet's coffin, when the weight of the iron was neutralized. The Captain & I agree that on these occasions three cheers should be given as the coffin disappears, not in levity, but as a genial hearty fare-thee-well wherever you are. Did a fair day's work, about 60 I should think. Made a bad miss in the evening. *Polynia* has 2050 seals, worse than us.

Tuesday April 13th

Boiled Beef day again (Tuesday – Teugh-day – Tough-day – Boiled Beef day). The worst dinner in the week except Friday. Lay to on account of the gale all day. Had the gloves down in the stokehole in the evening and some fine boxing. No seals.

[DRAWING "In Memoriam Andr. Milne April 11th 1880"]

Wednesday April 14th

Knocking along among the ice under sail and canvas picking up seals. Made a

---

113. Plum duff is nearly the last thing someone with intussusception should eat. This common dessert aboard whalers and other vessels long at sea was a stiff pudding made of water, flour, molasses, and raisins (the "plums" of the dish's name) – heavy ballast even for healthy men with good digestions. Fortunately for Conan Doyle's medical reputation, he appears to have not been present when the plum duff was served to poor Milne.

` 114. According to Dr. Robert Katz: "By this point, Milne probably suffered irreversible damage (infarction) of part of the intestine with perforation of the bowel and leakage of bowel contents into the peritoneal cavity (peritonitis). A plate of plum duff would have just made things worse and death shortly after eating this is not surprising. Most conditions like this require surgical intervention, clearly impossible for Conan Doyle to provide on a whaler in the middle of the ocean. Regardless of diagnosis, there was not much that he could have done for this poor fellow."

good day's work, about 80 I should think, bringing us up to 2450 about. Stood on the fo'c'sle head all day and reported progress. Rather cold work, had a shot or two tho'. Someone told me that in the South Seas when a man died the first comer got his property, and that when a man fell overboard you might see half a dozen standing by the hatchways to run down for the plunder whenever he was drowned.

Thursday April 15th

Beautifully fine day but we did a poor day's work, about 46 I think. Assisted in shooting 2 bladders. They took five balls each. A pretty little bird with a red tuft on its head, rather larger than a sparrow came and fluttered about the boats. No one had ever seen one like it before. Rather a long beak, feet not webbed, white underneath, with a "pea-wheet – pea-wheet." A sort of Snowflake.[115] Georgey Grant got his trousers torn by a young Sea Elephant in the evening.

Friday April 16th

Steamed hard to the North West all day to see if we could see anything of the seals. Failed in seeing many, and only picked up half a dozen. Jack Buchan shot a hawk in the evening which the Captain with his eagle eye discerned upon a hummock, and detected even at that great distance to be a hawk. About 18 inches high with beautifully speckled plumage.

[3 DRAWINGS: "My idea of a hawk," "The Captain's idea of a hawk," AND "The prey the Captain's hawk is looking out for."]

Saturday April 17th

Nothing doing all day. Only half a dozen seals again. We are steering South now with the *Iceberg*, a Norwegian. If we could only make it thirty tons I wd be satisfied. We have about 28 now I think. 26° of frost today. Had singing in the evening in the mates' berth.

[DRAWING 'Saturday's Night at Sea. April 17th/80.']

I began a poem on tobacco which I think is not bad. I never can finish them. Ce n'est que la <u>derniére</u> pas qui coute.[116]

---

115. Small Arctic bird better known as the Snow Bunting.
116. "It is only the final result that counts."

Sunday April 18th

A snowy drizzly kind of a day. Shot a seal in the morning off the bows; it was just sticking its head over the water. Saw two large sea birds, "Burgomasters" they are called. Went to a Methodist meeting in the evening conducted by Johnny McLeod the engineer, he read a sermon from an evangelical magazine and then we sang a hymn together. Argued afterwards with him.

Monday April 19th

Started stuffing our hawk this morning, or rather skinning it, for that is all I can do having no wires. I opened the stomach, then got out the legs to the knees and the humeri, and then inverted the whole body through the hole, cleaning out the brain, and removing everything except the skull. The result was satisfactory. We got a few bladders today, and are going North now to the old sealing. The Captain seems not to like the look of the ice at all.

[DRAWING "A Snap Shot"]

Tuesday April 20th

Nothing doing all day. Didn't take a single seal. Sailed and steamed to the North East. 72.30 today. Cleaned a couple of seal's flippers for tobacco pouches, rubbed alum all over our hawk's skin.

Wednesday April 21st

Absolutely nothing to do except grumble, so we did that. A most disagreeable day with a nasty cross sea and swell. No seals and nothing but misery. Felt seedy all day. Was knocked out of bed at 1 AM to see a man forwards with palpitations of the heart. That didn't improve my temper.

Thursday April 22nd

A heavy swell still on. Took about 13 of which I shot two. Bad but better than yesterday. Thick fog. Got a newly pupped seal, it seems rather late in the season for that. I have shot hitherto about 15 seals. I intend to count them after this.

Friday April 23rd

Did rather better today taking 36 seals. I made a bag of 11, that is 26 altogether. The shooting was uncommonly bad on the whole. Looks like a gale this evening. Captain saw another hawk. It is an extraordinary thing that we have not fallen in with a bear yet. Captain saw a meteoric stone fall into the water once within a hundred yards of the ship. The Magnetic Pole is in King William's Land[117] Lat 69°. There is another for South Pole, a thing that I never knew before.

[DRAWING 'Our Evening Exercise']

Saturday April 24th

We have been steaming North West all day. Saw a fine flock of Eider ducks, the males are black and white, the females bronze with a green head. Picked up 17 more young seals. I think we are not very far from the old ones. Had a pleasant evening in the mates berth. No shooting today. Sparred in the morning. Have a tip to teach Jimmy.[118]

Was talking to Hulton, one of our best harpooners, about zoological curiosities. He says that during a gale between Quebec and Liverpool he saw two fish lying on the surface of the water. They were about 60 feet long and spotted all over, exactly like leopards. An unknown species. The Captain fell in with another species in Lat 68° the hide of which was so thick that no harpoon would pierce it. Here is my list of Northern Whales.

Right Whale	Proper Greenland whale. Yield 10–20 tons of oil. Bone sells at £1000 a ton. Value of one is £1500–2000. Found in far North between the ice fields.
Finner Whale	Found in every sea in hundreds. Are longer and stronger than the Right Whale, but very worthless. Some 120 feet long. Razor backed. Spout two jets, proper whale has only one.
Bottlenose Whales	Found South of the ice, & round Iceland. Only 30 feet long. Give a ton of oil (£80). Skin valuable.
White Whale (Beluga)	Found everywhere, including Westminster Aquarium. Chiefly at mouths of American rivers. Oil valuable. 16 ft long.
Black Whale	a rare variety. Captain has only seen one. Valuable. Americans get them sometimes off North Cape.
Hulton's Whale	(Balaena variagatum)
Capt Gray's Whale	(Balaena ironsidum)

---

117. Part of Canada's Arctic regions.
118. Presumably his friend James Ryan back in Edinburgh.

Sunday April 25th

Got among a current of young bladders in the morning and took 22. I did good shooting before dinner, hitting seven in eight shots from the bows. Shot one after dinner and missed two which was poor. We have 2502 now. Saw one old seal. Boxed with Stewart and sang hymns with Johnny. Drew a fine picture of young sealing. Saw a good parody.[119]

> "Oh the wild Rhymes he made,
> Small poets wondered
> To see in the 'Light Brigade'
> 'Hundred' and 'Thundered'"

Monday April 26th

Sailing N and NW all day trying for old seals. They lie on the points of heavy ice stretching out into the sea, but you never know exactly where you can come across them; you must just coast along the heavy Greenland ice until you find them. We are 74° North today. Took one young seal yesterday, and saw several. We have nobler game in view. Boxed in evening. Challenged Stewart to run a hundred yards. I understand the sealing business thoroughly now.

[FOLDED-OVER DRAWING 'All hands over the bows – young sealing. 1880.']
[DRAWING 'Plan of Greenland Seal fishing']

Tuesday April 27th

Steaming N and NW all day. We have been among young bay ice and are trying to make the heavy where we may expect seals. Looks as if we were not far off towards evening. Did nothing all day. The skin of my hawk is just ruined. Drew Milne's funeral again at night for his brother who is aboard.

Wednesday April 28th

Made the heavy Greenland ice early in the morning and when I came on deck after breakfast it was stretching along the whole horizon. Heavier ice than any I have seen yet. The effects of the Arctic refraction are very curious. Here are two views at the distance of a mile and close up.

[TWO DRAWINGS OF 'Heavy ice close up' and 'Heavy ice at a mile']

119. Of Tennyson's "The Charge of the Light Brigade" (1854).

Saw marks of large herd of seals on the ice. A few in water steaming Northward. Was close to the *Victor* of Dundee in the evening. Beggar has no right to be there. Have the best prospects for tomorrow.

Thursday April 29th

Our prospects have not been realized, for although we saw a few schools of seals in the water, we have not reached their headquarters though we have been steaming North all day. Capt Davidson of the *Victor* boarded us this morning – a poor specimen of a man, hairy also. Was our mate once, and had the reputation of being a sulky beggar. The effects of refraction were extraordinary tonight, many pieces of ice appearing high up in the air with the sky above and below them. *Victor* steamed after us all day. We are not far from the seals I'll bet. Saw many Snowbirds about which is a good sign. In Latitude 75°.11.

Friday April 30th

Morning broke very inauspiciously with a Southerly wind and a hazy sky. We are beginning to feel a bit downhearted as our sealing should be begun by this time. Steamed North East after the haze rose, water was like a lake with a great deal of bay ice and numerous loons and petrels. Just before tea we saw a point of heavy ice ahead, and hope to find the seals at the other side of it. I am rather doubtful as we have seen none in the water as yet. The night is very nearly as bright as the day now, I can read Chambers' Journal at midnight easily. Served out grog this evening as tomorrow is the first of May. The ice looks well for the whaling. 10. PM As I thought there are no signs of seals upon the ice, so we have come to the conclusion that we are probably to the North of them. 'Mine too, Mammy." They generally shave newcomers on the first of May and a boatsteerer told me this evening that I was to be a victim, but I told him they would have to call two watches to do it.[120] [SCRIBBLED OVER: I am fairly sure I could chase the whole of this watch off the deck if I was properly rushed [?] and I feel just in the humour for a row. I will wait till midnight and see what sort of a job they will make of it.]

---

120. An ordeal, says Sutherland's *Whaling Years*, p. 41: "In the early hours of May Day the greenhands were taken on deck and subjected to a tortuous initiation ceremony. Senior crewmen, masked and dressed in strange seal-skin robes, blindfolded the new hands, scrubbed their faces until they bled, and then shaved them with a rusty saw-like razor. A vile cocktail was then forced down their throats as a toast to 'Their Oceanic Majesties, Mr and Mrs Neptune.' The bizarre affair was thus concluded, and the new men were accepted as members of the 'Brotherhood of Greenland Freemen,' the 'Honourable Fraternity of the Blubber Hunter.'" Buchan's *Peterhead Whaling Trade* has a similarly gruesome account. Conan Doyle may have avoided it, but readers of Sherlock Holmes will be reminded of the Scowrers' initiation ceremony in *The Valley of Fear*.

Saturday May 1st

In the morning there was a heavy swell on and our prospects were of the darkest. Before dinner however there was a change, for we saw a young bladder on the ice, and shortly afterwards a considerable school of seals in the water going before the wind. The Captain and all of us were rather gloomy at dinner time, but the moment he mounted the crow's nest after dinner down came the welcome shout "Call all hands." I was in the second mate's boat and we lowered away about 4.30 PM. It was a fine sight to see the seven long whale boats springing through the blue water as we made for the ice. The seals seemed a good deal whiter than when I saw them last. Mate fired and missed, and not a shot could I get, as he took the boatsteerer with his club and left me to find my way, and test the ice with the butt end of my rifle as best I could. Which was rather scurvy conduct. It was dangerous work on the ice, as I could see no one, and twice only just saved myself from falling in when I should in all probability have been drowned. It is more dangerous work than the young sealing, for the sea undermines these lumps of heavy ice, so that when you think you are perfectly safe on a large piece, the whole thing may crumble thro'. Never got a shot the whole time. Mate slew one. Most miserable work, the worst boat of the lot. If he could shoot as well as I, or I could walk on ice as well as him, we would have had a different tale to tell. However it is jolly to get to work again. Seals were soon frightened off the piece. Went off to the Westward. Here is our day's pickings

Buchan's Boat	14	
Rennie's Boat	13	
Carner's Boat	10	
Colin's Boat	10	Total 69.
Mathieson's Boat	10	
Hulton's Boat	9	
McKenzie's Boat	2	
Cane's Boat	1	

Sunday May 2nd

Showers, heavy ice, snow and wind all conspiring to ruin us. We steamed North in the teeth of a gale all day persevering manfully. In the evening the Captain came down from the masthead almost in despair, and pointed out a "blink" in the sky showing heavy ice ahead. "If the seals are not there," he said "I must turn South again." We steamed along and then to my delight after tea "All

Hands" were suddenly called. A considerable body of seals were in sight but as others were seen coming on to the ice, it was thought advisable to leave them tonight & attack them tomorrow. The great thing is to try and get a turn at them before the *Victor* sees what we are doing. This Dundee ship is the only one in sight. The Captain of her is a sumph.[121] Turned in early for an early rise.

Monday May 3rd

Boats lowered away about 6 AM. The moment they were down the *Victor*, who is about five miles off, turned & steamed furiously towards us. I went with Peter McKenzie, the last of the harpooners. We call our boat the "mob." It is manned by all the rag tag and bobtail of the ship, but I think has as good a crew as any. There is Peter harpooner, Jack Coull steerer,[122] the Doctor, Steward, Second Engineer, and Keith the oldest man in the ship. We were the last to leave the ship as the boats were dropped one here, one there. The ice was very heavy and good. At first we had a bad berth and only shot 2 seals, but we poked and pushed our way by sheer hard work through the ice and got into a fine bay lined with seals. Peter and I sprang out with our guns and wriggled our way along the ice, while the crew crept after us to skin what we shot. I saw Peter shoot two, and then I floored one. Then I got behind a hummock and shot nine, five all in a line on the edge of one piece. I was just thinking we would make a good bag and had shot another, while I could hear the ring of Peter's rifle a hundred yards off, when in came the *Victor*'s boats, pell mell all in a heap right at the back of us. The men sprang out, rushing across the ice firing without aiming, jumping up on the top of hummocks, shouting, and making the most fearful mess of it. They scared the seals and spoilt our work and their own too. I don't suppose they got fifty seals all together. Our boat had 27 and the united total of our morning's was 234, Hulton heading the poll with 68, and Cane having only 8.

---

121. Scottish term for a blockhead or dunce.

122. In "Life on a Greenland Whaler," Conan Doyle described this man anecdotally, and called him "our handsome outlaw" in identifying him as the boatsteerer: "There was only one man on board who belonged neither to Scotland nor to Shetland, and he was the mystery of the ship. He was a tall, swarthy, dark-eyed man, with blue-black hair and beard, singularly handsome features, and a curious reckless swing of his shoulders when he walked. It was rumoured that he came from the South of England, and that he had fled thence to avoid the law. He made friends with no one, and spoke very seldom, but he was one of the smartest seamen in the ship. I could believe from his appearance that his temper was Satanic, and that the crime for which he was hiding may have been a bloody one. Only once he gave us a glimpse of his hidden fires. The cook – a very burly, powerful man – the little mate was only assistant – had a private store of rum, and treated himself so liberally to it that for three successive days the dinner of the crew was ruined. On the third day our silent outlaw approached the cook with a brass saucepan in his hand. He said nothing, but struck the man such a frightful blow that his head flew through the bottom, and the sides of the pan were left dangling round his neck. The half-drunken and half-stunned cook talked of fighting, but he was soon made to feel that the sympathy of the ship was against him, so he reeled back grumbling to his duties, while the avenger relapsed into his usual moody indifference. We heard no further complaints about the cooking." This appears to be John Coull, forty-one years old in 1880, a seaman on Greenland whalers out of Peterhead – though according to Scottish census records, Coull was born in Peterhead.

Our Captain lowered the ensign 3 times to the *Victor* as an ironical "Thanks for your politeness." The moment the boats were aboard we set off at a great rate as the Captain saw a fresh and a larger batch of seals. We had a mouthful of dinner it being 2 P.M. When we came on deck after it we found the *Victor* had already landed her boats, so choosing another spot off we went. There was an enormous body of seals but very shy, so that we had to make long shots. We got 28 this time and the total came to 287 or so. Altogether today we got 540, a splendid day's work, about 11 tons of oil. Felt tired as I had been pulling and crawling on my face all day. Captain sees another patch of seals for tomorrow.

Tuesday May 4th

At it at 6 AM again. Boats lowered away and dropped here and there as usual. Peter and I got behind a hummock and shot 7 each, when the Captain saw he had not got into the thick of them, so he hoisted the Jack as a signal to the boats to return, he took the first five that came up, including our noble selves in tow, and away he steamed at full speed right past the *Victor*'s boats and dropped us in among a fine patch. It was an energetic and sensible action. I suppose he towed us about 15 miles or so. We made good use of our chances then and shot away hard. The "mob" distinguished itself killing 41, Buchan was best with 75, then Colin with 51, Carner 42, we 41 and the rest very poor. We took 275 and did not lower away again. A Norwegian ship the *Diana* came in on our flank but did not get very many. One of their boats came alongside ours and we asked them if they had seen the *Eclipse* to the North, they said they had, but I doubt if they understood us. *Victor* had its men out all night last night – a very short sighted policy.

Wednesday May 5th

Steamed to the NE. Open water round us. Hardly expected any seals.

[DRAWING 'Five Bulls at a hundred yards']

All hands were called however just before dinner. The *Diana* got the better of us rather, having all her boats in the heart of the pack before we lowered away. The seals were lying very thick but not over any great extent of ice. Our fellows muddled it completely, each being anxious to get the best position and beat the others. The seals were finally scared off after we had taken 71. Captain seems displeased and quite right too.

	May 1	3	4	5	14	15	
Colin	10	35	51	8	7	16	69
Cane	1	36	20	2	13		540
Carner	10	61	42	11	2		275
Hulton	9	112	11	6	12	14	71
Buchan	14	87	75	18	20		119
Rennie	13	68	26	7	2		32
Mathieson	10	47	10	11	2		1106
McKenzie	2	55	41	8	5		2502
	69	540	275	71	63	32	3608

Scrimmidge   56 ✓

Game Bag of the "Hope." Voyage 1880

					My Bag
April 3	760 Young Seals.		57 Old Seals.	1 Old Seal
4	450 Young Seals.		10 Old Seals.	
5	400 Y. S.			
6	270 Y. S.	6 Bladders.	57 Old Seals.	2 Blad.
7	133 Y. S.			
8	30 Y. S.			
9	50 Y. S.			
10	72 Y. S.			2 Seals
11				
12				
13				
14	80 Y. S.			2 Seals.
15	46 Y. S.	2 Bladders.		2 Seals. 1 Blad.
16	6 Y. S.	a Hawk.		
17	10 Y. S.	2 Seals.		
18	10 Y. S.			1 Seal.
19	6 Y. S.			
20				
21				
22	13 Y. S.			2 Seals.
23	36 Y. S.			11 Seals.
April 24	17 Y. S.			
25	22 Y. S.			8 Seals.
26				
27				
28				

29					
30					
May 1	69	Old Seals			
2					
3	540	Old Seals		27 Seals.	
4	275	Old Seals		10 Seals.	
5	71	Old Seals.			
6					
7					
8					
9					
10					
11					
12					
13					
14	119	Old Seals			
15	32	Old Seals.			
16					
17	6	Old Seals.			
18					
19					
20					
21					
22					
23					
24					
25					
26					
27					
28					
29					
30	2	Ground Seals			
31	1	Flaw Rat.			
June 1	1	Bladdernose.			
2	4	Roaches.	7 Loons.		
3					
4					
5	1	Bladdernose			
6	1	Narwhal.	2 Rare Ducks		
7					
8	1	Roach.	1 Loon.		
9	1	Roach.	6 Snowbirds		
10	1	Kittiwake.	1 Maulie.	3 Loons	
11	a	Whiting			
12	a	Bear			
13					
14					

15				
16				
17				
18	a Bear & 2 Cubs.			
19				
20	a Bear			
21				
22				
23				
24				
25				
26	a Greenland whale			
27				
28				
29				
30	a Burgomaster –	Snowbird –	5 Loons.	1 Flaw Rat.[123]

May 6th Thursday

The Captain has come to the conclusion that we had better go South a bit for seals, and let the *Diana* and *Victor* go on to the North. We are in 77.20 today, and expected to see the west coast of Spitzbergen. All hands are making off, there is a heavy swell and no signs of seals as yet.

[FOLDED-OVER DRAWING 'Pull, boys, pull!']

Steamed SW all day but saw no seals. We have about 50 tons now.

[DRAWING 'Process of making off in Sealing']

Friday May 7th

The *Diana* seeing us lying to yesterday night thought we saw something splendid so she came down at a fearful rate to share the booty under steam and canvas. After burning 30/ worth of coals it began to dawn upon the Norwegian mind that the whole thing was a "do" and a sell, so with a howl of disgust it flitted off again, to Iceland we believe. Under sail all day to N.E. Saw some schools in the water.

Saturday May 8th

Steaming NW. *Victor* in sight going in the same direction. He follows us as a jackal follows a lion. Ice all marked with seals. A beautiful day. Gave out tobacco

---

123. Or floe rat: throughout his log, Conan Doyle uses the terms flaw and floe interchangeably. A flaw rat was the smallest variety of seal, he explains later in his entry for May 12th.

and sugar in the evening. Was amused by a sailor's auction. Manson Turville a Shetlander was auctioneer & was particularly eloquent about a very dilapidated and seedy old coat of his which he wanted to palm off. "Going at five plugs of tobacco, at five plugs! Nobody bid any more? A coat warranted to keep out anything under 190 degrees of frost – no advance on five plugs? Gentlemen! Gentlemen! Five plugs and a half. Thank you, sir! Going at five and a half! The figure of a beaver will be found on one side of the lining and a rattlesnake on the other. Not sold but given away! Going, going now or never. Gentlemen, now's your chance for a bargain. Gone." I bought a pair of sealskin trousers from Henry Polson.[124]

[DRAWING 'Not Sold but Given away.']

Sunday May 9th

Why are seals the most holy of animals? Because it is mentioned in the Apocalypse that at the last day an angel shall open six of them in heaven. None of them to be seen today. The thing is growing monotonous as Mark Twain said when the cow fell down the chimney for the third time while he was composing poetry below.[125] A cloudy day. Have been reading Scoresby's book on whaling. Some of the anecdotes are too big to be swallowed at a gulp, they need chewing. He saw a whale caught in the bight of a rope that held another whale fast. Saw a man go a quarter of a mile on a live whale's back. However on the whole it is an eminently readable book, and very accurate as far as I can judge! Nothing all day. Was down in the harpooners' berth in the evening, conversation ran on zoology, murders, executions and ironclads. Steaming to the Northwest.

Monday May 10

We are down in 73.20 now or only just to the N of the place where we were young sealing. We are going North again I am glad to say. No seals. Served out coffee and tea in the morning. Glass had a tremendous fall after tea, and it came on thick rain and wind. I hope we are going to have a bit of a hurricane. Anything to wake us up. A codfish has been brought up through the pumps in a case of a big leak.

Tuesday May 11

A heavy gale during the night, and nearly all day. We hardly feel more than

---

124. Manson Turville is unidentified; Henry Polson was a middle-aged harpooner and cooper.
125. From Mark Twain's *The Innocents Abroad*, 1869.

the force of the wind however, as the ice forms fine natural harbours. Running North all day. It is too bad this – after we began our old sealing so well too. However this is a trade of ups and downs, and we must wait for the swing of the pendulum. Old Peter got a nasty cut over the eye tonight from a rope, and seemed to think he was blinded, but I set him right again. Misery & desolation.

Wednesday May 12

A most beautiful day. Blue sky which is rare up here as the sky is usually rather peasoupish. A good many seals in the water but none on the ice. As clear at midnight as during the day. A "finner" whale was seen spouting near the ship after dinner, but I was asleep myself at the time. I would have liked to have seen it. Balaena Physalis is its scientific name, and it is the swiftest, strongest, biggest and most worthless of the whale tribe, so hunting them is rather a losing game. However there is a regular finner fishery. They are worth about £120 each, and our whale about £1500, so we are on the right side of the bargain. Played Catch the Ten[126] in the mates' berth for love. The last time I had a card in my hand was at Greenhill Place. Saw a "flaw rat" today swimming round the ship. It is the smallest variety of seals. Captain's idea for the cure of baldness. Pick hairs out of another man's head by the roots. Then bore little holes in your head and plant them. He dreamed it.

Thursday May 13th

I hear from the engine room that Mr. McLeod, our chief engineer, has done me the honour to read my private log every morning, and make satirical comments upon it at table, and among his own firemen. Now I would as soon that he read my private letters as my journal, in fact a good deal sooner, and it is just one of those things which I won't stand from any man. If any man meddles with my private business I know how to deal with him. I am only astonished that a man professing religious principles should act with such a want, I won't say of gentlemanly honour, but of common honesty. If he does it after this warning he shall answer for it to me. A sensible man might be trusted, but a man who will talk about my prejudices against boiled beef &c. in the engine room must be suppressed. I hope this may meet his eye in the morning.

---

126. Calling it "Scratch the Ten," Albert Hasting Markham, p. 73, describes it as "rather a noisy game ... whose particular feature appears to be a grand fight as to which shall obtain possession of the ten of trumps." It was not peculiar to whalers, as he believed, but its popularity among them may account for it also being known as Scotch Whist. (Conan Doyle's subsequent comment confirms that the "Captain" he played in Lerwick was not a card game, but something else like "The Captain's Mistress.")

Saw a "finner" whale today. I had no idea of the size and sleekness of the brutes before. His blast looked like a puff of white smoke. He was a good quarter of a mile from the ship but when he dived I could see every fin. A most enormous creature. Prospects look brighter this morning as seals have been seen on the ice, and a good many in the water. Cane came running down about 11 PM to say that he saw a good strip on the ice.

Friday May 14th

Boats lowered away about 9 AM. Seals would not lie at all though. They have come up all the way from the Labrador coast, and are nothing but skin and hair after their month's traveling. Our boat was one of the highest with 5. We took 63 in the boats & then came aboard. The harpooners were sent over the bows to attack the remainder of the pack and killed 56 more making a total of 119. It is pleasant to get started again. Cane was frightened by an enormous walrus with a head like a barrel coming up beside him while he was flinching a seal on the ice.[127] He fired 4 shots into it, but it only seemed amused, and swam away.

[DRAWING 'Poking the "mob" boat through heavy ice. May 14th 1880.']

Saturday May 15th

Sent away two boats in the morning, Colin's and Hulton's, to a small patch of seals. Took 32 of them. Steamed and sailed North afterwards but saw no more blubber. The time is rapidly approaching now when we must coil our whale lines and go North.

Unless we fall among seals in the next two days, we must give it up. We think the *Eclipse* and the *Windward* are North already, we have seen nothing of them since more than a month ago, when they ran down to Iceland for the bottle-nosing. Reading Scoresby's journal of his discoveries between Lat 69° & 74° on the coast of east Greenland. The last Danish settlements on that coast are a very curious problem. He found no trace of them.[128]

127. Flinching: slicing the skin and fat off the carcass of a seal or whale.
128. "One of the most romantic questions in history," claimed Conan Doyle in *Through the Magic Door*; "I have strained my eyes to see across the ice-floes the Greenland coast at the point (or near it) where the old Eyrbyggia must have stood.... the Scandinavian city, founded by colonists from Iceland, which grew to be a considerable place, so much so that they sent to Denmark for a bishop. That bishop, coming out to his see, found that he was unable to reach it on account of a climatic change which had brought down the ice and filled the strait between Iceland and Greenland. From that day to this no one has been able to say what has become of these old Scandinavians .... It would be strange if some Nansen or Peary were to stumble upon the remains of the old colony, and find possibly in that antiseptic atmosphere a complete mummy of some bygone civilisation." Conan Doyle raised it in his Portsmouth Literary & Scientific Society talk, said the *Hampshire Post*: "it was a matter of speculation whether these highly cultured people had intermarried with the savage aborigines (a very unlikely event), whether they had become extinct (also unlikely, seeing that they were hardy men), or whether they remained, an isolated people with strange old-world customs and learning." He revisited it nine years later in "The Glamour of the Arctic," calling it "one of those interesting historical questions, like the fate of those Vandals who

Sunday May 16th

No seals. Lat 76.33 at noon. Banging away to the North as hard as we can go. Port wine. Old Cooper tumbled down the hatch and broke his arm nearly.

Monday May 17th

A beautiful day. Steamed to the North all day. Lat 77° Longt 5 East. About 100 miles west of Spitzbergen. Got 6 from a small patch of seals after dinner. They are getting very thin. The ones we have captured lately we consider to be, not Greenland seals, but seals from the Labrador coast which after their month's traveling could hardly be expected to be in prime condition. Rigged up the harpoon guns this morning. Fearfully cumbrous things working on a swivel with a pull of 28 lbs.

[DRAWING OF A HARPOON GUN]

Has to be let off by pulling at a string. Carries a harpoon about 30 yards with some accuracy. Base about 1½ inches.

[DRAWING OF A LOADED HARPOON GUN MOUNTED ON A BOAT, SIGNED 'Capt. J. G.']

Tuesday May 18th

Cleaned out the boats and made all straight for whaling. During dinner a sail was seen to the N.W, which turned out on closer inspection to be the *Windward*, which we thought was South with the *Eclipse* at the bottlenose fishing. We hauled our yards aback and waited for him. Murray came aboard with a very dismal tale to tell. After the young sealing he had been too ambitious to content himself with the modest work that we had stuck to, picking up half a ton a day or so, but he had run right away North at once to Spitzbergen after whales, not taking a single old seal. The result of his ambition is that we have about 52 tons now, and he about 28. He has been here three weeks and never seen a fish. He gives a most discouraging account of the whole thing, and will, I think, go away after the bladders. A heavy gale blew from the SW during the evening, a most awkward direction for us.

Wednesday & Thursday May 19 & 20

Blowing a hard gale both days. We are tacking and turning between the ice

were driven by Belisarius into the interior of Africa, which are far better unsolved. When we know everything about this earth, the romance and the poetry will all have been wiped away from it. There is nothing so artistic as a haze."

and Spitzbergen. We can make out the *Windward* in the lulls, sometimes ahead, sometimes astern. Sea running very high, and sky as dark as possible. Took a sea aboard on Wednesday, giving the watch a fine ducking. My old foe, Toothache, has it seems followed me all the way from Scotland, and been hiding about the ship the whole voyage. Yesterday it came out from its concealment and said "Ah, mine enemy, and have I found thee?", whereupon it seized hold of me by one of my incisors and twinged it so, that my whole face is distorted today. (Addison).[129] On Thursday we saw the wild bleak coast of Spitzbergen breaking through the rifts of the storm. A great line of huge black perpendicular crags running up to several thousand feet, as black as coal but all seamed with lines of snow. A horrible looking place. We were thinking of running in and anchoring in King's Bay, but the chart was mislaid. Toothache.

Friday May 21st

Spitzbergen still in sight about 50 miles to the North East. A complete lull in the wind but the sky very dark, and a heavy swell on, from which we think we will have a change of wind. *Windward* went South, and in the evening we saw the *Eclipse* coming up in the distance. As we had not seen her for a month we were anxious to know what she had done. The Captain boarded her and after three hours came back with the news that she had been down to Iceland and had managed to capture 32 bottlenose whales, a very large take. They yield on an average a ton apiece, and as Captain David had also as many young and old seals as we, he has managed to beat us so far. 90 tons I believe he has got. This wind has done us terrible injury by packing all the ice up close, and destroying all the bights or bays in it in which whales are usually found. However we must just keep up our peckers,[130] and hope for the best.

Saturday May 22nd

A heavy swell all day. I come of age today.[131] Rather a funny sort of place to do it in, only 600 miles or so from the North Pole.

[FOLDED-OVER DRAWING 'The Hope in a gale off Spitzbergen']

---

129. Despite a seeming reference to Joseph Addison (1672–1719), he seems to paraphrase First Kings 21:20: "And Ahab said to Elijah, Hast thou found me, O mine enemy? And he answered, I have found thee: because thou hast sold thyself to work evil in the sight of the Lord."

130. A 19th-century British expression referring to the nose, akin to "keep your chin up."

131. Conan Doyle was born in Edinburgh on May 22, 1859.

Had rather a doleful evening on my birthday, as I was very seedy for some reason or another. The Captain was very kind and made me bolt two enormous mustard emetics which made me feel as if I had swallowed Mount Vesuvius, but did me a lot of good. *Eclipse* sailing near us all day. Ice is sadly damaged.

Sunday May 23d

Plum Duff day again – a fine day, the swell all gone. Sailing in to the West again down the tight ice. The Captain and I have been making most villainous parodies of Jean Ingelow's "Sparrows Build"[132]

~~"When Sparrows build & the L~~
"When 'Burgies' build their Greenland nest,
    My spirit groans and pines
For I know there are seals in the Nor'Nor'West
    But its time to 'coil our lines.'
Far down in the South the 'Bladders' lie
    But the Devil a one near me,
And the 'Unis' are sticking their horns on high
    As they plunge & play in the sea

Chorus
But oh the Whale, and the right, right whale!
    And the whale we all love so
Is there never a 'bight' in the Greenland 'tight'
    Where a ~~whale has room to~~ 12 foot whale can blow.

Thou didst set thy foot on the ship and fare,
To that ~~sad~~ cold and lonely shore,
Thou wert sad for the seals were all skin & hair,
And they came from Labrador.
And 'Meg' he came and scared away
Some twenty mile at the least,
And how could we tell where the 'Flappies' lay,
With a great bay flaw to the east.
                         Chorus
We shall never again sail back in May,
As we oft before have done
Or take four thousand 'young' in a day

---

132. The first verse of "When Sparrows Build" by Jean Ingelow (1820–97), a popular English poet, indicates its appeal to Captain Gray and Conan Doyle: "When sparrows build, and the leaves break forth, / My old sorrow wakes and cries, / For I know there is dawn in the far, far north, / And a scarlet sun doth rise; / Like a scarlet fleece the snow-field spreads, / And the icy founts run free, / And the bergs begin to bow their heads, / And plunge, and sail in the sea."

And go home with two hundred ton,
We shall never be full with seals alone
For all our work and toil,
But we'll never say die while whales yield bone,
And 'Bladders' give up their oil

             Chorus.

Monday May 24th

Another fine day. We are going to have a little luck at last I hope. 6 P.M. No, we are not though. We are certainly awfully unlucky this year. A strong wind has set in from the East and is packing up a nice little bight which was forming, and playing the deuce with our prospects. Colin says we have a Jonah aboard. *Eclipse* near us. Got our harpoon guns stuck up.

Tuesday May 25th

Worser and worserer. Wind still blowing from the East and murdering the ice. All hands disgusted. *Eclipse* set sail for the South but seemed to think better of it and came back again. Horrible!

Wednesday May 26th

A fine day but the ice is ruined. Wormed our way through it as best we could. I was smoking my afternoon pipe on the quarterdeck when there was a cry of "A bear – close to the ship." Captain was at the masthead and sang out at once to "Lower away the Quarter Boat." I ran down for my shooting iron[133] and succeeded in getting a seat in the boat. I could see the bear – a great brute – looking quite tawny against the white snow, and running very fast in a direction parallel to the ship. Then he crouched down in a hole of water about a couple of hundred yards off, and hid with just his nose above surface. Mathieson was harpooner of the boat and we pulled off, but had to make a bit of a circuit to get through the ice. We lost sight of him, and when we saw him again he was standing with his forepaws on the top of a hummock and his head in the air, staring at us and sniffing. We were within shot then but we thought he would let us get nearer, so we bent to our oars. But some associations connected with boats seemed to dawn on his obtuse intellect, for he suddenly got off the hummock and we lost sight of him. Then we saw the signal hoisted for the boat to come aboard,

---

133. An Americanism turning up in Mayne Reid's work Conan Doyle read as a boy: avid for tales about the American West, he drew on them in his first Sherlock Holmes tale, *A Study in Scarlet*.

and spied Bruin travelling over the ice at a great rate, and a long way off, so we had to give it up as a bad job. Wind still Easterly.

Average for the Old Sealing

Buchan	14	+	87	+	75	+	18	+	20	=	214		
Colin	10	+	85	+	51	+	8	+	7	=	161		
Hulton	9	+	112	+	11	+	6	+	12	=	150		
Carner	10	+	61	+	42	+	11	+	2	=	126		
McKenzie	2	+	55	+	41	+	8	+	5	=	111	(mob)	
Rennie	13	+	68	+	26	+	7	+	2	=	116		
Mathieson	10	+	47	+	10	+	11	+	2	=	80		
Cane	1	+	36	+	20	+	2	+	13	=	72		

Scramble 55
Young Sealing 141
Total. 1216 Old Seals

[DRAWING 'The Bear we did <u>not</u> shoot']

Thursday May 27th

Ice began to close round us fast in the morning, and we had to steam our way out to the open sea as best we could, to save ourselves from being beset or nipped. Had a difficult job to get out. *Eclipse* kept in our wake. Captain went on board her to dinner and stayed till about eight. I drew, slept, played draughts and boxed while he was away. We are going off to 80° North Latitude to see what is up there, right up to the Northern Barrier in fact.[134] The terror of the seas up here is an animal which is called the Swordfish, but is not a swordfish at all. It is one of the whale tribe with a long snout like a mackerel, and great pointed teeth the whole length of its jaws. It attains the length of 25 feet, and is distinguished by a high curved dorsal fin. It feeds on the largest sharks, on seals and on whales. Yule of the *Esquimeaux* took six whales the other year in the Straits, which actually came and cowered under the ship for protection, because one of these monsters was in the vicinity. The Captain tells me that he was in the crow's nest one day when he saw a great hubbub ahead of the ship. On examination with the glass he made out that it was an enormous sea elephant which was sitting on a piece of ice very little larger than itself. In the water round it were half a dozen of these bloodthirsty fish, which were striking the poor creature with their long fins, trying to knock him off his perch when they would have made short work of

---

134. Where the ice made further progress in the direction of the North Pole impossible.

him. As the ship came up the Captain says he never shall forget the look which the poor seal cast towards it with its big eyes, and suddenly taking an enormous bound off the bit of ice, it squattered along the surface of the water, and took such a leap towards the ship's side, that its head was above the taffrail, and it very nearly gained the deck. A boat was lowered, when the great 12 foot creature climbed into it, and was knocked on the head. Balls had to be fired into the fish to keep them from attacking the boat, they were so riled at the disappearance of their prey.[135]

[DRAWING OF 'The Greenland Sword fish' SIGNED Capt J. G.]

Friday May 28th

Steaming North and North East all day, in company with the *Eclipse*. It is clear we are not going to have any whales in May, and we can only hope for the best in June. Thick fog in the evening and we had to blow our steam whistle and fire guns for several hours before we could find the *Eclipse* which was also screaming loudly. Took a bang at some loons on the ice at a long range with my rifle, since signals were the order of the day, but although I hit the piece and knocked the snow all over them I slew none.

Saturday May 29th

[DRAWING 'The Loon or Lesser Auk.']

The less said about Saturday the better. Let Saturday sink into oblivion. Nothing doing. Fog in the evening. Lat 79.10 North at noon. Played cards in evening.

Sunday May 30th

Captain David came aboard in the morning and expressed great dissatisfaction at the state of the ice, in fact he said he had never seen it worse. Dr. Walker came afterwards with a logbook for me which the Captain very kindly sent. In

---

135. Conan Doyle was not long home when controversy about the Swordfish (*orca gladiator*, the grampus or killer whale) erupted in *The Scotsman*. On September 15th Lord Archibald Campbell, recently crossed from Quebec, described a battle he had witnessed when a whale was attacked by both a thresher shark and some manner of swordfish. The following day T.G. of Edinburgh, though unable to identify the whale, was confident that even such formidable predators as a thresher shark and swordfish could not prevail unless it had been sick. Conan Doyle wrote next, suggesting T.G. mistook the swordfish for something else: "That frightful creature 'the swordfish,' the scourge of the northern seas, is a very different fish to the poor 'lang-nebbit' brute [Scots for long-nosed] which T.G. confuses it with. *Orca gladiator*, or the killing grampus, is a ferocious species of whale and received the soubriquet of 'swordfish' from the old Greenland whalers on account of its long and high dorsal fin. It is furnished with rows of formidable teeth, and being extremely fast in the water, and larger than any shark, it may fairly lay claim to be the most formidable of the denizens of the deep." He provided an instance or two of its prowess against whales, identifying the one in question as the hunchback, "a not uncommon fish in such a latitude .... I have seen them among herring south of Jan Mayen Island in large numbers." (T.G. was not pleased to be corrected, but a Westminster Aquarium naturalist named J. Abrahams, on the 22nd, gave it as his learned opinion that the whale was most likely *Rorqualis borealis*.)

the morning we espied two objects swimming near the ice, which the Captain made out to be two ground seals, a rare variety, nearly as large as bladdernoses. We lowered away a boat and after an exciting chase, and an exhibition of bad shooting on the part of the harpooner, we nailed them both. They were a female and [a] young one, the former about 8 ft. 6. By the way, talking of bladdernoses, Colin killed one once which measured 14 feet long. It charged the boat and nearly bit the harpoon gun in two. We hope we may have a turn in the luck now after this small capture. By the way one of the most interesting things in Arctic Zoology was the capture last year of a large albatross by Capt David, in 80° North Latitude. Where did the breed come from? It looks as if the temperature of the Pole was semitropical.

Monday May 31st

Dreamed about whales in the Caledonian Canal[136] and how frightened we were lest some of the barges or horses would scare them. There were 17 of them, all bottlenoses under a bridge. A very curious dream. The wind this morning is WNW which is excellent, and the water is of a greenish hue which is excellenter.

[FOLDED-OVER DRAWING 'Swordfish in pursuit of a school of seals']
[DRAWING 'A capture']

By the way I haven't half exhausted my curious dream. While we were away killing the whales under the canal bridge I heard it strike two o'clock, and it suddenly came into my head that my final professional was to have begun at one. Horrified, I abandoned the whales and rushed to the University. The janitor refused me admission to the examination room, and after a desperate hand to hand struggle he ejected me. Even then I did not wake, but dreamed that someone handed me out a paper to see what the questions were like. There were four questions but I forget the two middle ones. The first was "Where is the water ten miles deep near Berlin?" The last was headed NAVIGATION, and the question was this word for word, "If a man and his wife and a horse were in a boat, how could the wife get the man and the horse out of the boat without swamping it?" I grumbled very much at these questions, and said it was not fair to introduce Navigation into a medical examination. Then I determined to send the paper to Captain Gray and get him to answer it, and then at last I woke up. Certainly the most connected dream, as well as the most vivid I ever had.[137]

136. Completed in 1822, the Caledonian Canal connects Scotland's east coast at Inverness with its west coast at Corpach, though only a third of its length is man-made.
137. Earlier (April 1) the young medical student had remarked at his not being examined that day for the first time

This evening our foretop yard came down with a run owing to the breaking of a shackle, and smashed the halliards [halyards]. We put up a spare spar and made all right again within four hours, a fine bit of seamanship. Captain has gone up to the nest and I am writing this before the cheery cabin fire. I hear the hammering on deck as they do up the broken yard, and just outside the door the Steward is remarking in a really first class tenor that 'At midnight on the sea-leas – Her bright smile haunts me still.'[138] It seems to haunt him at midnight, and then he employs the odd 23 hours in commenting upon the fact. Captain David was on board in the evening, and lent me a pamphlet on whales.[139] I was experimenting on the 'maulies' in the evening. I took 4 pieces of bread and soaked them, one in strychnine, one in carbolic acid, one in sulphate of zinc, and one in turpentine. Then I threw them over to the birds to see which would work quickest, but to my horror an old patriarch stepped forward and swallowed the whole four pieces, and strange to say he didn't seem a bit the worse.

Tuesday June 1st

I trust that we may have better luck this month, than last. We can see the Northern Barrier along the whole horizon. *Eclipse* is getting up steam and I suppose we are going to have a look at the bight in Lat 78° out of which we were driven, and then we will run away to Scoresby Sound on the Liverpool Coast[140] if nothing turns up. Water is full of animalculae and olive green in colour.

> Balaena mystractus!
>   Balaena mysticetus!
> If we were animalculae
>   You wouldn't take long to eat us.

Captain says he has seen whales spouting so thick that it looked like the smoke of a large town. A very good description. By the way the water rose 8 degrees yesterday, from which we think we are in the Gulf Stream. Passed a piece of

---

in three years. He had striven to compress his studies for financial reasons, but now his suspending them to go off on a seven-month adventure seems to be at least a subconscious cause of anxiety. "I have more than once received important information through my dreams," remarks a character in a future tale of terror of his, "The Leather Funnel."

138. By W. T. Wrigton and J. E. Carpenter of Macon, Georgia, in 1864. "He had a very beautiful and sympathetic tenor voice," Conan Doyle said in "Life on a Greenland Whaler," "and many an hour have I listened to it with its accompaniment of rattling plates and jingling knives as he cleaned up the dishes in his pantry. He knew a great store of pathetic and sentimental songs, and it is only when you have not seen a woman's face for six months that you realize what sentiment means. When Jack trilled out 'Her Bright Smile Haunts Me Still,' or 'Wait for Me at Heaven's Gate, Sweet Belle Mahone,' he filled us all with a vague, sweet discontent, which comes back to me now as I think of it. As to his boxing, he practiced with me every day, and became a formidable opponent – especially when there was a sea on, when, with his more experienced sea-legs, he could come charging down with the heel of the ship."

139. Likely the one he and John Gray had penned themselves in 1874, *Report on New Whaling Grounds in the South Seas* (Aberdeen: D. Chalmers), which drew attention to reports of right whales in the Antarctic's Ross and Weddell Seas, leading to an expedition in 1892 to investigate.

140. Not Liverpool, England, but a region on the eastern coast of Greenland north of Iceland.

a fir tree floating in the water. It has come many thousands of miles, drifting down the Obi or Yenesei rivers in Siberia and so into the Arctic seas by the NW current.

Saw two bladdernoses in the evening but only got one on account of bad shooting. Hoped to get away and shoot roaches in the evening but there weren't any. Buchan shot four in the morning.

Wednesday June 2

Plying West and South under canvas. Captain suffering from Ablubberomnia. Very cold, as cold as it was in April. My hair is coming out and I am getting prematurely aged. Read a good story that a doctor was buried in the middle of a large churchyard, and a professional brother suggested as an epitaph "Si monumentum quaeris, circumspice."[141] Very witty, I think. It is very disheartening to be kept off the whaling banks like this by the ice. As the Stewart says "it makes a lad inclined to jump up, and never come down again." Sydney Smith said of Jeffrey "His body is too small to cover his mind. Jeffrey's intellect is always indecently exposed."[142] Very clever too. Saw the marks of a large bear in the evening, also a bladdernose in the water. Things look rather more hopeful this evening, as we have made considerable way to the Westward, and are close to the whalebanks.

Thursday June 3d

Very cold again, a great hoar frost is on. Strong wind from the North. We are in the most promising place we have seen yet and if the wind holds we ought to catch some of the minnows we are after. Came on a fog in the evening. I have my cod line overboard baited with pork but have not had a bite. About 50 sail of Russians come to Spitzbergen this month to hunt cod, so there must be some knocking about. Had the bag net out tonight and towed it to see if there was any food. Brought up a most beautiful Clio or Sea Snail, a couple of inches long, looking like some weird little fairy. I have stuck him in a pickle bottle and christened him "John Thomas." I hope he will live, we have put some butter and pork into his house. Saw a good many narwhals knocking about, one very large

---

141. "If you seek his monument, look around."

142. Francis Jeffrey (1773–1850) won a permanent place in criticism's history by declaring that Byron's poems "would not do." Conan Doyle used him in his first attempted novel *The Narrative of John Smith* as an example of criticism's often mistaken judgments. Sydney Smith (1771–1845) co-founded *The Edinburgh Review* with Jeffrey.

one, almost snow white and quite 15 feet, ricocheted past the stern, making the peculiar grunt they give when they rise. Also saw some beautiful medusae.[143]

Friday June 4th

John Thomas is in an awful passion. We left the pickle bottle far from the fire, and as there are 11 degrees of frost it froze up and John has caught cold. He is sitting in a corner with his tail in his mouth, just as a sulky baby sticks its thumb into its potato box. I have drawn John's attention to the butter & pork and he took a hurried breakfast, but seems to have business of importance down at the bottom of the bottle. He's thinking perhaps of

> Where his rude shell by the Gulf Stream lay,
> There were his little Sea Snails all at play,
> There their Amoeboid mother, he their sire
> Butchered to make a whale's holiday.

[DRAWINGS OF 'John Thomas coming up for his breakfast (sign of life)' AND 'A small friend of John's']

Just after one o'clock I was standing on deck talking to Andrew Hulton about the general bleakness of affairs. He is one of our best harpooners. I happened to ask "By the way, Andrew, when a man does see a whale I suppose he never sings out 'There she blows' as put in books." He said "Oh they cry 'There's a fish' or anything that comes into their head, but there's Colin going up to the masthead, so I must go on the bridge." Up he went on the bridge and the moment he got there he bellowed out "There's a fish!" There was a rush for the boats by the watch but the Captain put a stop to it. "Do things quietly," he said "and man the boats when I give the word." We could see two blasts ahead among the ice, and I caught a glimpse of the back of one of the great creatures as he dived. We lowered away the boats of the watch and afterwards four others, six boats in all, and our hopes ran high but were alas doomed to disappointment. Two other fish appeared, and the four went off to the W N W. The boats kept after them (heavy ice) for four hours, but it was no go, something seemed to have scared them. However it is something to be on their trail. We see a flaw water[144] ahead and hope we are going to have a fine time of it in there. The *Eclipse* was also after two fish but lost them owing to the unskilfulness of their second mate who lost his post in consequence.

---

143. A jellyfish plentiful in Arctic waters.
144. A space of open water created by a rent in the ice.

Saturday June 6th [5th]

John is well and hearty. Saw a great many narwhals today, but none of what we want. Kept a lookout on the bridge from breakfast to dinner. We saw a large sea elephant on the ice about noon, and Andrew Hulton & I went away and shot it. About 9 feet long and very fat. We opened its stomach and found it contained a very large assortment of cuttlefish. Captain went aboard the *Eclipse* in the evening. The guano here is blood red, and has a curious effect. Plenty of birds about. Wind coming round to the South. It is a most exciting business, the tension on the nerves is very great.

Sunday June 6th

John was up before me and took a heavy breakfast. He is now gyrating round the top of his bottle surveying his new kingdom apparently and meditating a map. I put him in a bucket every evening where he wanders fancy free for an hour or two. Wind is round to SW, I am glad to say, it was S SW yesterday. We may see fish any moment now, the water is a peculiar dark grey green. I thought I smelled a fish yesterday from the deck. You can often smell the greasy smell long before you see them. Aaron our Shetland boy, the son of old Peter the prophet, was in the crew of the boat that visited the *Eclipse* yesterday. When he came back I heard him go straight to his father and begin with

"Father, Peter Shane's been treaming!"
(Peter Shane is the rival prophet of the *Eclipse*)
"Ay, boy. What?"
"Peter Shane's had a tream, Father."
"And what did he tream, boy?"
"He treamed he saw them killing cows on poard the *Hope*."
"Oh a good tream, boy, a good tream. That means that the *Hope* will have the first fish. A very good tream."

So we still have some hopes.

Saw a large cuttlefish under the surface, and a good many medusae and clios. About 3 PM word came that the *Eclipse* boats were away. They were several hours after their fish but finally they were recalled by the hoisting of the bucket.[145] About 6 PM Adam Carner saw a blast a long way off from the masthead. Four boats were sent off in pursuit, but failed even to catch a glimpse of the whale. Jack Buchan, who by the way started in his shirt and trousers just as he tumbled

---

145. A globe of hoops covered with canvas, used to signal whale-boats to return to their ship.

out of bed, nailed a narwhal or Sea Unicorn about 13 feet long, with a horn of 2 feet. The harpoon cut its throat most beautifully. It was towed by the four boats and hoisted aboard. Beautifully speckled with black and grey. After flinching it we opened the stomach which we found to be full of a very large shrimp, which I take to be the "Mountebank" shrimp, and with lots of cuttlefish. It had two distinct sets of parasites upon it, one like a long thin worm in the drum of its ear, the other seed-like at the root of the horn.

Two very rare ducks were seen behind the ship this evening. The Captain went off himself in a boat and nailed them both with a right and left barrel. No one on board has seen the species before. They have a yellowish beak with an orange callosity stretching up in a curve from the base of the beak towards the eyes. They are rather larger than our ducks, dark brown on the head, white on the neck, dark brown on the back, lighter silvery brown on the breast. All the plumage very soft & delicate.

[DRAWING 'Towing home the Narwhal']
[DRAWING 'The Narwhal itself']

Monday June 7th

No fish seen today though the Captain thought he had a glimpse of one in the evening. I went aboard the *Eclipse* after tea to get some arsenical soap to preserve our ducks with. Captain David says he thinks they are King Eider ducks, a very rare bird. Captain David came back with me and stayed an hour. He was after three fish yesterday but got none. I caught a petrel by flinging a lead with a bit of string attached over its head, when the string warped round it and I hauled it in. It looked confoundedly astonished. I let it away again. Wind North & North East. Blowing hard in the morning

Tuesday June 8th

Steamed a bit in the morning. Sun shining brightly. Secured the ship by an anchor to a piece of floating ice, and whistled for a change of wind. Sent away three expeditions after narwhals but without any success. Went away birdshooting but only got two shots killing a roach and maiming a loon. Peppered a flaw rat but it got away. Grant saw the steps of an Arctic fox on the ice. Had a pleasant day on the whole. Captain says if I will load my own cartridges I may blaze away until all is blue. Made sail again in the evening. Played 'Nap'[146] in the engine room. Almost dead calm.

---

146. According to Hoyle, Nap, or Napoleon, was a card game for two to six players, with quite a few variants including one called "Wellington and Blucher."

Wednesday June 9th

We were forced to come out towards the open sea again today on account of changes in the ice. *Eclipse* and we moored ourselves on to one piece of ice in the evening. Captain David & Dr Walker came aboard us about ten PM and stayed until two. They shot a very large bear upon the ice today. It was sitting munching away at the head of a narwhal which it had dragged on to the ice, while a great shark was wiring into the tail which hung over into the water. How the bear got a narwhal onto the ice is a mystery. Went away at two o'clock in the morning to shoot birds. They were very scarce however and I was only enabled to get a roach and 6 snowbird. Saw a large bladdernose but were unable to get a shot at it. Came back at 4 PM.

Thursday June 10th

Still trying hard to get in to where we know the whales are lying. Made some progress under steam and then anchored with the *Eclipse* to an iceberg. Shot a kittiwake and a loon off the deck, and then got two more loons while picking up the first. Amused myself in the afternoon by catching petrels by flinging a lead over the heads of them, and warping the string round their wings, something like the South American "bolas." By the way when I shot a roach the other day a great maulie seized it the moment it fell and regardless of the shouts of the boat's crew, and my frantic howls, proceeded to bear it away, but I shied a boat's stretcher at it and scared it off.

[DRAWING 'Maulie Stealing our Roach']

> John Thomas
> died on the 8th of June, regretted
> by a large circle of acquaintances

He was a right thinking and high minded Clio, distinguished among his brother sea snails for his mental activity as well as for physical perfection. He never looked down upon his smaller associates because they were protozoa while he could fairly lay claim to belong to the high family of the Echinodermata or Annulosa. He never taunted them with their want of a water vascular system, nor did he parade his own double chain of ganglia. He was a modest and unassuming blob of protoplasm, and could get through more fat pork in a day than many an animal of far higher pretensions. His parents were both swallowed by a whale in his infancy, so that what education he had was due entirely to his own industry and observation. He has gone the way of all flesh so peace be to his molecules.

Zoological List of Whaling Voyage

INVERTEBRATA
  I Protozoa
        Any number in whale's food
  II Infusoria
        Rice food.
  III Annulosa

Common Louse (on a Shetlander)	Clio Helicina
Shrimp (common)	Horn louse of Narwhal
Clio Borealis (John Thomas)	Ear louse of Narwhal
Shrimp "Mountebank"	Whale louse (Ocina)

  IV Echinodermata
        Medusa gulius   78.40 N.
        Medusa ——? 78.40 N.
        Flask-shaped medusa 78.5 N.
  V Mollusca
        Sepia ——   78.40 N.

VERTEBRATA
  I AVES
        Arctic Petrel or "Maulie" (Procellaria Glacialis)
        Foolish Guillemot or "Loon" (Colymbus Troile)
        Roach (Arca Alle)
        Doveca (Colymbus Grylle)
        Burgomaster (Larus Glaucus)
        Kittiwake (Larus Rissa)
        Snowbird (Larus Eburneus)
        Snow Bunting (Emberiza Nivalis)
        Redpoll (Fringilla Linaria) 75 N.
        Puffin or "Tammy Norie" (Alca Artica) 78 N.
        Boatswain (Larus Crepidatus) 78.12 N.
        Iceland hawk (Falco Icelandicus) 73.40 N.
        Great White Owl (Stryx scandiaca) 71 N.
        Great Tern or Sea Swallow (Sterna Hirundo) 78.18 N.
        Brent Geese (Anas Bernicla) 78 N.
        Eider duck (Anas Mollissima)
        Sea Gull (Larus Communis)
        Shag (Lerwick)
        Duck (Calvo?) Very rare. King Eider. 78.50 N.

Arctic Starling   78.6 N.
Sandpiper   75.30 N.
Arctic Gull   69° N.

PISCES

Flaw fish (rather like a whiting)   78.40 N.
Silver Fish   78.12 N.
Herring   69°.
Squalus Greenlandicus or Greenland Shark

MAMMALIA

Horsesaddle Seal (Phoca Vitulina)
Bladdernosed Seal
Flaw Rat
Ground Seal   79 N.
Walrus (Trichechus Rosmarus)   77.30 N.
Whitefaced Seal
Fresh Water Seal   (78 N.)
Orca Gladiator. The Greenland Swordfish. Lat 69 N,
Bottlenosed Whale (Delphinus Deductor)   63 N.
Razorback Whale (Balaena Physalis)
Narwhal (Monodon Monoceros)
Right Whale (Balaena Mysticetus)
Balaena Musculus (Hunchback whale)   68° N.
Ursus Maritimus or Polar bear
Canis Lagopus or Arctic Fox

Additional birds seen on passage
Solan Goose
Stienchuck
Stormy Petrel
Black Back Gull
Sparrow Hawk
Mallet

Friday June 11th

We made a few miles in the right direction. The *Eclipse* shot two bears this
morning. About one o'clock a fish came up near the *Eclipse* but was not captured.
We made fast to an iceberg in the evening. Caught a curious fish today, the first
I have seen in Greenland. It looked rather like a whiting, but was not one. Jack
Williamson one of our hands got a terrible blow from the wheel. It exposed the
bone of his skull for about 5 inches. Stitched it up and sent him to bed. Steward,
the boatsteerer and I were walking on the ice in the evening and both distinctly

saw the blast of a fish about half a mile off among the pack ice. It could not however be reached by boats.

Saturday June 12th

The ice is shutting rather than opening. Hope deferred maketh the heart sick. Men shot a bear off the side about eight o'clock. I was asleep and so missed the fun. Stomach was full of seal oil but he was very thin [several words illegible] shots. Went aboard the *Eclipse* at dinner time with the Captain. Had a pleasant feed and chat. Captain David seems far from despairing. Strong wind from W and SW – ought to do us good. Ice began to close round us so rapidly that we had to steam out 30 miles or so to prevent being nipped or beset. Had a difficult job to get out as it was. The sea this morning was actually swarming with narwhals.

Sunday June 13th

Got a fine opening towards the Westward and worked in again about as far as we came out, going W & SW. Saw nothing but one seal in the water the whole time. Need about 20 miles north to take us into whaling ground. "Thou art so near and yet so far." It does seem hard after our penetrating impenetrable packs, and leaving forty miles of shifting heavy ice between us and the sea, exposing ourselves to every danger of storm and flood, and putting ourselves in the way of losing our ship and ourselves, or of being beset and wintering out, that we sh[oul]d be no better off than the half hearted beggars who shun the whole concern, and go South after small game. It is a shame and a sin, and can't last long. The *Eclipse* and ourselves are the last of ten generations of daring Arctic seamen, the breed has deteriorated and we are the sole survivors of the men who used to harry Greenland from the 80 to the 72, and here we are stuck in the mud & helpless. It would make a saint swear.

Monday June 14th

Thick fog in the morning. Blew foghorns but got no answer from the *Eclipse*. Jack Williamson, the man with the head, is doing very well. Things look as bad as they can be and worse. I hope we will go and hunt bladdernoses instead of persevering at this. The whales are only 20 miles off but an impassable barrier of ice intervenes, and the wind is such as to pack the ice firmer together, rather than to open it out. We want wind from the W, NW or NWW and we are getting it all from the South. ο ποποι ποποι![147] 53 tons! Such is life!

---

147. "Oh strange! oh shame!" – an expression of pain in Homer's *Iliad* and other Greek literature.

Tuesday June 15th

The only difference in the weather is that the fog is thicker and the wind more utterly odious and depraved. However we are at the bottom of our woes for nothing could make our situation worse, so "there's an end on 't" as old Sam Johnson used to say. Captain went aboard the *Eclipse* at dinner time. I do hope we'll go and slay bladdernoses or bottlenoses or any other animal who has some peculiarity about its nose, and carries blubber on its carcase. Browsed over Boswell all day.

Wednesday June 16th

The *Eclipse* lowered away after a whale about 8 AM and pursued it until noon, but did not get it. The fact is we are not upon the grounds, and any we see are stragglers on the march, and not stopping to feed. Wind Westerly, so far so good. Calm in the evening, sea looks like quicksilver, the whole place covered with narwhals, great brutes 15 & 16 feet long. You hear their peculiar "Sumph!" in every direction. I saw one pass like a great flickering white ghost underneath the keel. Reading "Tristram Shandy," a coarse book but a very clever one.[148]

Thursday June 17th

An eventful day – for the *Eclipse* at any rate. In the morning about 10 AM Colin saw a whale from the crow's nest about five miles off while the *Eclipse* had her boats after another. We made sail and reached up towards Colin's fish but did not see it again until about 1 PM when it suddenly appeared within 50 yards of the ship, accompanied by another one. The two were gamboling and frisking in the water like a couple of lambs. We lowered away four boats, Colin's, Carner's, Rennie's and Peter's which all pulled up for a piece of ice where the fish were likely to reappear. They came up there near Rennie's boat, but he unfortunately is not a man of much decision of character, and he hesitated to fire into the nearest fish for fear of scaring the other, which was turned eye on to him. The fish separated, one disappearing and the other leading the boats a most exciting chase to windward. From the deck I could see its blast rising apparently just in front of the boats, and its great tail waving in the air, but our men could never get quite within shot of it. The *Eclipse*, seeing the way the whale was heading, came round that way and dropped two boats in front of it. The whale came up

---

148. *The Life and Opinions of Tristram Shandy, Gentleman*, by Laurence Sterne, 1767, notorious for its bawdy humour. Samuel Johnson, whose *Life* Conan Doyle had been browsing, said: "Nothing odd will do long. *Tristram Shandy* did not last." – prompting Schopenhauer to comment that "Sterne is worth a thousand pedants and commonplace fellows like Dr. J."

in front of one, the second mate's, and in a moment we had the mortification of seeing the boat's Jack flying, as a sign they were fast to our fish. It is hard to see a thousand pounds slip through your fingers so. They killed the fish during dinner and had it aboard before 8 PM. Rennie got a fine wigging from the Captain when he came aboard.

[DRAWING ILLUSTRATING THE STORY ABOVE]

After dinner we saw a large bear on a point of ice apparently in a great state of excitement, probably due to the smell of the whale's blood. I got off in the boat with Mathieson to kill it. We got out on the ice, a great flaw many miles across,[149] with our rifles, and could see the bear poking about among some hummocks about 40 yards away from us. Suddenly he caught sight of us and came for us at a great speed, running across the pieces of ice towards us, and lifting up his forefeet as he ran in a very feline way. Mathieson and I were kneeling down on the ice, and I intended not to fire until the brute was right on the top of us. Mathieson however let blaze when it was about 15 yards off, and just grazed its head. It turned and began trotting away from us, and as it only presented its stern I was compelled to put my bullet into that. It was wounded but went off across the ice at a great rate, and we never saw it more.

Saw a "boatswain" gull today. Row in the mates' berth in the evening.

Friday June 18th

Eclipse struck another fish during the night and had it aboard before breakfast. Lucky dogs! Buchan shot a fine bear and two cubs during the night. By the way my bear of yesterday, when it had escaped some distance, got up on its hind legs on the top of a hummock like a dancing bear, to have a good look at us.

[DRAWING OF 'Our Bear']

I had no idea they did that in a state of nature. Cruised about all day in search of blubber but found none. Our boats and the *Eclipse* were after one fish at night but they never got a start.[150] Ice is closing round us and we are cut off from the

---

149. Enormous floating islands of ice broken off shore-line shelves and carried by the current. In *Greenland, the Adjacent Seas, and the North West Passage to the Pacific Ocean* (New York: James Eastburn, 1818, pp. 144–47), Bernard O'Reilly described flaw ice as "sometimes leagues in extent, invariably level, and covered with snow about ten inches deep." A ship had to be wary of its movement, he said, lest it be caught between a huge slab of flaw ice and other obstructions, and its hull crushed.

150. While night-time by the clock, in this season the sun does not set in such northern latitudes. "Night seems more orange-tinted and subdued than day, but there is no great difference," Conan Doyle remembered in "Life on a Greenland Whaler": "Some captains have been known to turn their hours right round out of caprice, with breakfast at night and supper at ten in the morning. There are your twenty-four hours, and you may carve them as you like. After a month or two the eyes grow weary of the eternal light, and you appreciate what a soothing thing our darkness is."

sea, so that unless there comes a change of wind, we may easily be beset. We are all very melancholy.

Saturday June 19th

Calm as a fishpond, water like quicksilver. A good many narwhals about. The ice is remaining stationary or thereabouts. No fish seen today at all. A shark was seen to come up alongside and nail a maulie out of the water. We wouldn't mind being unsuccessful if others shared the same fate, but it is maddening that the *Eclipse* should make £3000 while we have not made a penny. Our Captain is as good a fisherman as ever came to Greenland, there are no two opinions on board on that point, quite as good as his brother David, but somehow the luck seems to be with the others. They have seen from first to last about 14 fish to our 5.

Sunday June 20th

A large fish was seen during breakfast, but after a short pursuit it got among pack ice where it was impossible to follow it. It was very nearly within reach once. This is terrible, to see fish and not to get them.[151] No man who has not experienced it can imagine the intense excitement of whale fishing. The rarity of the animal, the difficulty attending any approach to its haunts, its extreme value, its strength, sagacity and size, all give it a charm. A large bear was shot during the night.

Ice closed round us during the day but relaxed towards evening. Captain David came aboard during the day, and our Captain went and had supper with him. One of his harpooners was attacked in his boat by a bear the other day when he had no rifle with him, but he banged the hard wad of the harpoon gun

---

151. This may be the origin of an anecdote about Colin McLean, the first mate, that Conan Doyle told in "Life on a Greenland Whaler": "His only fault was that he was a very hot-blooded man, and that a little would excite him to a frenzy. I have a vivid recollection of an evening which I spent in dragging him off the steward, who had imprudently made some criticism upon his way of attacking a whale which had escaped. Both men had had some rum, which had made the one argumentative and the other violent, and as we were all three seated in a space of about seven by four, it took some hard work to prevent bloodshed. Every now and then, just as I thought all danger was past, the steward would begin again with his fatuous, 'No offence, Colin, but all I says is that if you had been a bit quicker on the fush – ' I don't know how often this sentence was begun, but never once was it ended; for at the word 'fush,' Colin always seized him by the throat, and I Colin round the waist, and we struggled until we were all panting and exhausted. Then when the steward had recovered a little breath he would start that miserable sentence once more, and the 'fush' would be the signal for another encounter. I really believe that if I had not been there the mate would have hurt him, for he was quite the angriest man that I have ever seen."McLean wrote objecting to the *Strand*'s publisher, Sir George Newnes. Conan Doyle told the editor, H. Greenhough Smith, on January 15, 1897: "I am sorry Colin Maclean [sic] took my article amiss for he is an old shipmate and I have an esteem for him. I have therefore written him a letter to say so. I tell him however that when he denies the truth of it his memory fails him. I am in communication with Jack Lamb, the steward, the other actor in the affair, and I will get a corroboration from him. I also kept a diary every day on the *Hope*. It is stored at present but I have no doubt I should find the incident noted that very day. Lamb seems quite pleased with the allusions to him. I am very sorry Sir George should have been bothered about it." And on the 21st: "I have just had a letter from the Steward who entirely corroborates my yarn about the mate – so if you make any amends don't let there be any doubt as to the precise truth of it all."

through it, which was ingenious. That was nothing however to what one of our harpooners did a few years ago, which would be incredible if I did not know it to be true.[152] Buchan was sent to shoot a bear and two cubs on the ice, but they took to the water before he reached them. He passed the noose of a rope over the head of each, as they swam and snarled at him, and tied the ends of the three ropes to his thwarts. All the oars were then run in except the steer oar, and Buchan standing in the bows and banging them on the head with a boathook whenever they offered to turn, guided the boat right back to the ship, the bears towing it the whole way. The Stewart, who saw it, says the roaring could be heard a mile off. Some hopes for tomorrow.

Monday June 21st

Hopes not realized as usual. We are shut up in a small hole of water with nothing but great ice fields as far as we can see. If it goes on shutting up we will be beset and our chances of any success ruined. Colin bad with a sore throat. No fish about. Caught a beautiful Sea Lemon[153] yesterday floating on the surface but it died shortly after being brought on board. Saw a very curious sight at midnight, which you might come North a lifetime and never see. There were three distinct suns shining at the same time with equal brilliancy, and all begirt by beautiful rainbows, and with an inverted rainbow above the whole thing. A most wonderful spectacle.

[DRAWING 'A family party']

Tuesday June 22nd

An utterly uneventful day. We are still cooped up in the hole of water. Caught a rare & indeed undescribed medusa in the evening (Medusa Doilea Octostipata). Misery and Desolation.

Wednesday June 23d

Made our way out of our prison, by a most delicate and beautiful bit of maneuvering under steam. We came out about 60 miles among very heavy ice, the smallest piece of which could have crushed our ships like eggshells. Often we squeezed through between floes where the ship's sides were grinding against the

---

152. Conan Doyle has come a long way since May 9th when he questioned the credibility of some more believable stories in Scoresby's Arctic journals than this one!
153. A colourful variety of sea slug.

ice on each side. Steamed S and E. *Eclipse* went after a fish but never got a start. Glass falling rapidly.

Thursday June 24

Captain and I were knocked up at 6 AM by the mate's thrusting his tawny head into the cabin, singing out "A fish, sir," and disappearing up the cabin stairs like a lamplighter.[154] When we got on deck the mate's and Peter's boats were already on the seat of action where the fish had been seen.

[FOLDED-OVER DRAWING, 'Boats of the Eclipse and Hope in pursuit of two whales']

They caught another glimpse of it about a mile to leeward and pulled down towards it, but lost sight of it again. Meanwhile another very fine whale came up astern very near the ship and Hulton's and Rennie's boats were lowered away after it. The four boats pursued it for a couple of hours, when it began to blow extremely hard and a heavy sea arose so that some of the boats' headsheets were right under water. We had to get them on board, and let the whale alone. Blowing a very hard gale from the North East all day, 9 Wind Force.

Friday June 25th

Wind is still very strong though not as much so as before. Nothing seen during the day but a large finner whale, which is a bad sign. It is of no use to us, and it drives the right whale away from its feeding grounds. Played Nap in the evening. Wind only a fresh breeze now, have begun to steam to the North after the *Eclipse*.

Saturday June 26th

Things looked dark enough all day, but suddenly took a turn for the better. Nothing had been seen all day, and I had gone down to the cabin about 10 o'clock when I heard a sort of bustle on deck. Then I heard the Captain's voice from the masthead "Lower away the two waist boats!" I rushed into the mates' berth and gave the alarm, Colin was dressed but the second mate rushed on deck in his shirt with his trousers in his hand. When I got my head above the hatchway the very first thing I saw was the whale shooting its head out of the water and gamboling about at the other side of a large "sconce" piece of ice.[155] It was a beautiful night, with hardly a ripple on the deep green water. In jumped the crews into their boats, and the officers of the watch looked that their guns

154. "To run like a lamplighter," a then-common expression originating in the previous century.
155. To mariners, a sconce was a small water-washed iceberg.

were primed and ready, then they pushed off and the two long whale boats went crawling away on their wooden legs, one to one side of the bit of ice, the other to the other. Carner had hardly got up to the ice when the whale came up again about forty yards in front of the boat, throwing almost its whole body out of the water, and making the foam fly. There was a chorus of "Now, Adam – now's your chance!" from the line of eager watchers on the vessel's side. But Adam Carner, a grizzled and weatherbeaten harpooner, knows better. The whale's small eye is turned towards him and the boat lies as motionless as the ice behind it. But now "it has shifted, its tail is towards them – Pull, boys, pull!" Out shoots the boat from the ice – will the fish dive before he can get up to it? That is the question in every mind. He is nearing it, and it still lies motionless – nearer yet and nearer. Now he is standing up to his gun and has dropped his oar – "Three strokes, boys!" he says as he turns his quid in his cheek, and then there is a bang and a foaming of waters and a shouting, and then up goes the little red flag in Carner's boat and the whale line runs out merrily.

But the whale is far from taken because it is struck. The moment the Jack appeared in the boat there was a shout of "a fall" on board, and down went [the] other six boats to help the "fast" boat and nail the whale on its reappearance. I got into the mate's boat and away we pulled. Of course the whale may come up anywhere within the radius of the line it has taken out, which may amount to three or four miles, so our seven boats had to spread out over a considerable area. Five minutes passed – ten – fifteen – twenty, and after being away 25 minutes the brute came up between the second mate's and Rennie's boats, who fired into her and dispatched her. She proved to be a small fish, about 40 feet long, with 4 ft 1 in of bone, worth between £200 and £300. We gave three cheers and towed her to the ship. She was covered with very large crab lice which accounted for her erratic conduct in the water. Had her flinched and stowed away by 3 o'clock AM. I went to the crow's nest during the process to look out for another, which I didn't see. Went to bed at 6 AM and got up at 12.

[DRAWING OF THE WHALE'S TAIL MARKED '4 ft 1.' AND 'Adam Carner' UNDERNEATH]

Sunday June 27th

Not a thing to be seen all day, but about 4 AM Colin saw a very large fish in the distance. *Eclipse* lowered away 2 boats as well as we, and after getting one start they lost scent of her. She seems to have been a tremendous brute.

Monday June 28th

Nothing.

Tuesday June 29th

Master aboard the *Eclipse* last night until 2 AM. Lay to in a calm all day – nothing doing. Waiting for a chance of getting North, but ice looks bad. Got away to shoot at midnight and came back at 4 PM. Got a burgy, a snowbird and five loons. These burgies I think are the biggest of gulls after the albatross. They usually are about 5 feet from tip to tip.

Wednesday June 30th

Slept nearly all day after last night's exertion. Went aboard the *Eclipse* and had a great talking. Worked with a microscope aboard. Buchan shot a flaw rat. Hulton skinned my birds.

Thursday and Friday July 1 & 2

Lying to in a thick fog as we have been ever since last Monday. Nothing to chronicle except that Colin got a large narwhal early on Thursday morning. It took out a whole line (120 fathoms) and made a great fuss about being killed. A unicorn is worth about £10. The skin is of considerable value. I have a very decent Arctic museum by this time including a lot of interesting things. I have at present

1. An Esquimeaux pair of sealskin trousers
2. An Iceland falcon
3. My sealing knife and steel
4. Bone of bladdernose – shot myself
5. 2 bones of old seals
6. 2 foreflippers of young bladdernose
7. 2 foreflippers of a ground seal
8. a bear's head
9. Bristles of a bladder
10. a Burgomaster
11. Drums of whale's ears
12. 2 King Eider ducks

[FOUR DRAWINGS UNDER THE COLLECTIVE CAPTION 'Our First Fish.' – 'The Shot,' 'The Fast boat,' 'Waiting for her return,' AND 'Dead – Hurrah!']

13. Bits of lava found in King Eider duck.
14. (?) a Unicorn's horn   Added 2 Esquimeaux pouches
                                         a Kittiwake
                                         a bear's claw

Saturday July 3rd

It has cleared up and we are off to the happy hunting grounds. Sailed Nor' and Nor'West all day. Saw nothing but an extraordinarily small seal on the ice, about the size of a rabbit. It seemed as much amused at the appearance of the ship as we were at it. We are all despairing. The Stewart stuffed my ground seal's flippers very nicely with sawdust.

Sunday July 4th

Sailed North and then South again. Everything looks bleak and discouraging. No trace of whales or even of whale's food. A bladdernose was seen on the ice. A small bird something like a starling or thrush was flying round the ship. Saw a puffin. Have no heart to write much in the log. Reading Motley's "Rise of the Dutch Republic," a very fine history.[156]

Monday July 5th

Steamed into a flaw water, made fast in the evening. Saw several finner whales. *Eclipse* mistook one for a fish and lowered away his boats, which however were promptly recalled. Nothing else of interest seen during the day. Got some delicious fresh water off the salt water ice.

Tuesday July 6th

Dead calm. Sun beating down in a tropical manner though the temperature was only 36°. Tremendous glare from the ice flaws. Went aboard *Eclipse* in morning. Got away to shoot and nailed altogether 7 loons, a roach, a kittiwake, a snowbird, and a flaw rat. We had great fun securing the latter as our small shot did not suffice to kill it, and after a chase of at least half an hour, we harpooned it with boathooks when swimming under the water. We brought it aboard alive but the Captain humanely put it out of its misery. Got away again at night but found no game. A couple of Sea Swallows[157] played round the ship. Saw several finners. A very jolly day!

Wednesday July 7th

Steamed 20 or 30 miles South, and then on seeing indications of fish made sail. Captain David came aboard in the evening with his Engineer, and caught a rare

---

156. John Lothrop Motley (1814–77), an American historian and diplomat whose 1856 book *The Rise of the Dutch Republic* was translated into a number of languages and had a European vogue.
157. A variety of Arctic tern.

shrimp. Feel very much the better for yesterday's outing. Cooked Red Herrings for supper in a very scientific manner.

Thursday July 8th

Another memorable day. Sailed along the edge of a great flaw among very blue water with the *Eclipse* ahead of us. About 1 o'clock a whale, the first seen since Sunday week, came up close to Captain David's ship; he lowered away three boats after it and chased it until 4:15 PM when he succeeded in getting fast, and had her alongside by 8 PM and flinched by midnight. We dodged about hoping his fish would come in our direction when we would have been justified in securing it, but about 4 o'clock the welcome shout came from the masthead "there's another fish on the lee bow, sir!" Mathieson and Bob Cane lowered away after it, and were soon lost sight of among the ice, while we crowded along the side of the deck and waited. Then a groan went up as a large finner whale rose near the ship, for finners and "right" whales are deadly enemies, and we were afraid our quarry would be scared. I went down to the cabin to sooth[e] my disappointment with a smoke, when I heard the Captain yell "A fall! A fall!" from the masthead, which is the signal that the fish is struck. Up we tumbled many of the men only half dressed, and away went five long green whaleboats to the support of the "fast" boat and its companion. I got into Peter McKenzie's boat. We had hardly got clear of the ship's side when the boatsteerer announced that the fish was up, and was lashing out, fin and tail. Then we knew our work was cut out for us, for when a fish stays a very short time under water after being struck, it is reserving all its strength for a struggle with the boats. If the whale goes down and stays away half an hour it is generally so exhausted on returning to the surface that it falls an easy prey. The boats pulled up and Hulton and Carner fired into her and got fast. We were the next boat up and pulled on to [the] fish's head, where we lanced her deep in the neck. She gave a sort of shudder and started off at a great rate along the surface. Buchan pulled his boat on to her head as she advanced, by which senseless maneuver the prow of the boat was tilted up in the air, and finally the whole boat landed on the animal's back amid a shouting of men & snapping of oars and Buchan roaring "Pull! Sweep! Back! Hold water! Pull! What the devil are you feared at!" I said to Peter "Stand by to pick them up!" but they managed to shove the boat off without accident. The beast now made for a flaw and got beneath it, but soon reappeared when both Buchan and Rennie fired into her. She went under again, but once more reappeared right among our three boats and then the fun began.

We pulled on to her and in went our lances for five feet or so, the three boats tried to keep well at the side of her while she was always slewing around to bring her formidable tail to bear upon us. She nearly had our boat over once by coming up underneath it, but we managed to get it righted. Then we stood off from her while she went into her dying flurry, whipping the water into a foam, and then she slowly turned up on her back and died. We stood up in our boats and gave three hearty cheers. We towed her up to the ship and by 1 PM had her aboard. She was a fine fish, each lamina of whale bone being 9 foot 6 inches, yielding about 12 tons of oil. It is worth quite £1000 and has secured our voyage from being a failure. A large and very ugly shark came up and superintended the process of flinching the fish in spite of numerous knives passed through its body. I asked the Captain to let the Stewart and me go off in a boat and harpoon it, but he refused.

[DRAWING OF A WHALE'S TAIL WITH '9 foot 6 in. / Bob Cane.' INSIDE IT]

[FOLDED-OVER DRAWING, 'Whale dragging 2 fast boats through water']

[DRAWING 'Buchan's boat on the top of the fish']

Friday July 9th

Nothing doing. Everybody in a state of reaction after yesterday's capture.[158] The Captain says that Bob Cane managed the affair very well. Several finner whales were seen during the day. Beautiful sunshine.

It is a curious fact that the last whale the *Eclipse* captured only had one eye, and our friend of yesterday was also restricted to the same meagre allowance. The socket was perfectly empty. It may be that there is a breed of one-eyed Greenland whales.

Saturday July 10

We have made a mistake, I think, in heading North again. The South seems to me a greasier locality. Had the boat away after a bladder which we did not get. Played Euchre[159] four hours in the evening in the engine room. Query Who did Adam & Eve's children marry?

---

158. Writing his autobiography many years later, Conan Doyle remembered the *Hope* taking three whales instead of two, and made a fish-story of it in "Some Recollections of Sport." "I once had the best of an exchange of fishing stories, which does not sound like a testimonial to my veracity. It was in a Birmingham inn, and a commercial traveller was boasting of his successes. I ventured to back the weight of the last three fish which I had been concerned in catching against any day's take of his lifetime. He closed with the bet and quoted some large haul, 100 lbs. or more. 'Now, sir,' he asked triumphantly, 'what was the weight of your three fish?' 'Just over 200 tons,' I answered. 'Whales?' 'Yes, three Greenland whales.' 'I give you best,' he cried; but whether as a fisherman, or as a teller of fish stories, I am not sure."

159. An American card game for four players. It was new to Albert Hastings Markham in 1874 when the *Arctic* carried

"I am going to write to Leigh Smith of the *Eira* today and ask him for a photo of my noble self," Conan Doyle wrote to his mother that November: "I was taken you know with a distinguished group on the quarterdeck." Pictured in W.J.A. Grant's photograph besides Conan Doyle were Captain David Gray, Benjamin Leigh Smith, Captain John Gray, Dr. Robert Walker of the *Eclipse*, Dr. William Henry Neale of the *Eira*, and William Lofley, ice master aboard the *Eira*. (Courtesy of Hull Museums and Art Gallery.)

Sunday July 11

Got up late, and would have liked to have got up later, which is a sad moral state to be in. *Eclipse* got a bladder in the morning. Steamed to the Eastward with the *Eclipse* in the evening, by which proceeding we scared a whale. Saw many finners. About seven PM a steamer was reported about 20 miles to the Eastward. This is the first ship we have seen since the beginning of May. We steamed out and soon recognized it as the new discovery yacht of Leigh Smith's, the *Eira*.[160]

American seamen from the wrecked discovery vessel *Polaris*, remarking in his journal, p. 222: "The game of euchre has come into fashion, and the table is taken up by it."

160. As Conan Doyle says below, Benjamin Leigh Smith (1828–1913) was a gentleman of private means who spent freely on Arctic exploration between 1871 and 1882. The screw barquentine *Eira* had been built for him at Peterhead with advice from the Captains Gray.

He is going to try for the Pole if the ice is favourable, which it isn't,[161] and in any case to explore Franz Joseph Land and shoot deer. He is a private gentleman, a bachelor with £8000 a year, and has taken Spitzbergen to himself as a wife. When our ships came up we saluted the little *Eira* with ensigns and three cheers, which they returned. His men are in naval reserve uniform, officers in gold lace. The Captain went aboard her, while their doctor, Neale, the photographer,[162] the engineer and 2 mates all boarded us. The Captain came back about 1 PM and he and I with the *Eira*'s photographer & doctor made a night of it on champagne and sherry.[163] We had tinned salmon at 5 AM and turned in at 6.30.

Monday July 12

Anchored to a flaw with the *Eclipse* and *Eira*.[164] Unshipped our rudder which was damaged by ice. By the way we got our home news up to June 18th from the *Eira*. Got no letter from Edinburgh, but a very cheerful and pleasant one from Lottie. Surprised to hear that the Liberals have got in, disgusted also.[165] Invited aboard *Eclipse* to meet Leigh Smith and gang at dinner. Had mock turtle soup, fresh roast beef with potatoes, French beans and sauce, arrowroot pudding and pancakes with preserves, winding up on wine & cigars. A very respectable whaler's feed. Went aboard *Eira* afterwards. She is beautifully fitted up aft. Had more cigars and champagne. Got aboard at 12, after being photographed in a group. They came up by Jan Mayen and saw millions of bladdernoses in the 72.30, I hope we may come down for them.

Tuesday July 13

Steamed 20 miles South and stopped short, I don't know why. I fancy we might fill our ship now if we went straight down for those bladders, but we must go at once.

---

161. "On July 11th the *Eira* fell in with the *Eclipse* and *Hope*, whalers of Peterhead, commanded by those well-known and enterprising captains, David and John Gray," reported C. R. Markham, "The Voyage of the *Eira* and Mr. Leigh Smith's Arctic Discoveries in 1880," *Proceedings of the Royal Geographical Society*, March 1881, p. 130: "These experienced navigators reported that the ice was low down the coast of Spitzbergen, and that there was little chance of making progress in that direction.... Spitzbergen was sighted [by the *Eira*] on the 14th of July, but the report of the Grays was confirmed by the ice itself."

162. The photographer was W. J. A. Grant, whose work is an important visual record of Arctic exploration. See Willem F. J. Mörzer Bruyns, "Photography in the Arctic, 1876–84: the work of W. J. A. Grant," *Polar Record*, 2003.

163. Conan Doyle's fellow surgeon on the *Eira*, William Henry Neale, was three years older, and had received his M.B. in 1879. He was also on the *Eira*'s disaster-fraught voyage to Franz Josef Land the following year (see n166 below), and wrote a vivid account of it, "Castaways in the Frozen North," for *Wide World Magazine*, April 1898. Judging from his picture there, Dr. Neale could have passed for Dr. Watson; and like Watson was a product of the University of London's medical school. ("In the year 1878 I took my degree of Doctor of Medicine of the University of London," says Watson in *A Study in Scarlet*, the first tale.)

164. At 3 p.m. that day, says W. J. A. Grant in his diary of the *Eira*'s 1880 voyage, following some movement earlier: "at noon all three [ships] went N. a bit." His diary is appended to Markham's account referenced in n161 above.

165. Perhaps, but by the mid 1880s, he was an active Liberal Unionist, the party's wing dissenting from Prime Minister Gladstone's Home Rule policy for Ireland.

We are vacillating here too much, I think, however it is for the Captain to decide. The success of our voyage depends on these few days, it's our last chance of making a hit. *Eclipse* chased a bear and killed it in the water close to our ship. Left the *Eira* in the morning wishing them all success. A pleasant ship and a pleasant crew. She is black with a line of gold, about 200 tons burden, and 50 horsepower engines. I think I should like to be going out in her, although the prospect of seeing home again is pleasant. They left their letters with us.[166] Fog in the evening.

Wednesday July 14

Steamed and sailed South and Sou'West. *Eclipse* had their boats away in the evening, but it was only a finner which they mistook for a right whale. Foggy nearly all day. No news of any sort. Read our papers all day

Thursday July 15

Another uneventful day. Lounged about & smoked. Absolutely nothing to do. Very thick and foggy and all that is reprehensible. Saw a small scene of Goethe's 'Faust' which I am reading which I think is as vivid and weird as anything I ever read, far more gruesome than Shakespeare's witches.[167]

> Night – – – An open Plain
> Faust, Mephistopheles rushing past on black horses
> Faust – What are these hovering round the Ravenstone?
> Meph – I know not what they're shaping & preparing.
> Faust – They wave up – wave down. They bend – they stoop.
> Meph – A band of witches.
> Faust – They sprinkle and charm.
> Meph – On! On!

That is very awful, I think.

---

166. The *Eira* had left Peterhead June 19th, and did not return until October 12th. As Conan Doyle remarks, the ice was not favourable for a try at the Pole, so the *Eira* explored Franz Josef Land more thoroughly, then worked its way down Norway's coast, at one point running aground for several days. ("An Arctic Exploring Voyage," *The Scotsman*, October 12, 1880.) The *Eira* had worse luck in 1881: That August it was wrecked on Franz Josef Land's coast, and Smith and his crew spent ten hard months before being able to take to their longboats in June 1882 in search of rescue. In his 1883 Portsmouth Literary & Scientific Society talk, Conan Doyle paid tribute to them, the *Portsmouth Times* reported: "A more forlorn position than that of his crew and himself can hardly be imagined – wrecked upon the shores of a desolate Arctic land, far from all human aid, with the long winter creeping in upon them, and no possibility of escape.... they built a hut, and, having a good store of ammunition, shot a number of bears and walruses for their winter's larder. By a judicious enforcement of exercise, and an example of cheerfulness, the men's health was preserved during six [sic] months of darkness until the ice broke up in the spring, when they took to their boats, and were picked up eventually by the *Hope*, which had been sent out in search of them by Sir Allen Young."

167. Johann Wolfgang von Goethe's tragedy in two parts about a scholar who sells his soul to the Devil (Mephistopheles) for greater than earthly knowledge. The lines are the penultimate scene of Part I. The work appeared in its final form after Goethe's death in 1832. Below, a couple of days later, Conan Doyle discussed Goethe vs. Shakespeare with Captain David Gray.

Friday July 16

Still foggy. *Eclipse* had four boats away during the night, but without success. We do not know yet if he really saw a fish. Boarded him in the evening and learned that it most certainly was a fish, and that they very nearly secured her. They got near enough to touch her with a boathook as she swam under water. Captain David still seems to be very sanguine. Some of our stores are running short, we got some potatoes however from the *Eira*. Stayed till 2 AM on the *Eclipse*. Got some more papers. Many seals seen during the day in the water.

Saturday July 17

Absolutely nothing to do or to be done. It has been thick fog now for nearly a week. Steamed about 20 miles S and E. Captain David came aboard at night. We intend now to try the Liverpool Coast right down in west Greenland near the land Lat 73 N. Many heavy fish have been taken there late in the season by Capt David, notably in '69 when he took 12, striking the first on the 16th of July, and the last on the 4th of August. I remember I used to think that when a whaler saw a whale they always got it, as a matter of fact the average is about one fish in 20 attempts.

Sunday July 18

A little clearer today, not very much. Strong SSW breeze changing to a gale in the evening. Blew very hard all day, and all night. Dodged about under the lea of ice flaws to escape its fury. I wonder if Leigh Smith's vessel is caught in it. By the way I was photographed among a distinguished group on the quarterdeck of her, but as I was smoking a cigar during the operation I am afraid I'll be rather misty.

Monday July 19

Blowing a gale all day. Nothing to do and we did it.

Tuesday July 20

Cleared up a little and we did a good day's work steaming among great icefields about 40 miles S & W. If it keeps clear we may do something yet. There is an enormous accumulation of ice this year round the land, more than has ever been seen. We are 240 miles from it now, and the fields are almost continuous. I'm afraid we won't get in.

Wednesday July 21

Thick again, this fog is paralyzing. We are groping in the dark. Anchored to a flaw in the evening and the Captain and I went aboard the *Eclipse*.

[DRAWING 'Flinching a whale. July 18th 1880']

Had Nox Ambrosiana[168] from 8 to 2 AM. The Late Mr. Procter.[169] Captain David tells me of a fish he captured which had a lump the size and shape of a beehive on the fluke of its tail. He entered into a critical analysis of Goethe's Faust, comparing it with some of Shakespeare's plays, and showing where the former borrowed from the latter, so we are not altogether barbarous up here.

Thursday July 22

Still foggy and we continue anchored to the flaw. In the half decks in the morning discussing the loss of the *Atalanta*.[170] Saw 2 "boatswains," very rare birds, at a considerable distance over the flaw, and was going to hunt them but they absconded. Got a shot at a flaw rat's head about 50 yards off in the water, and blew it clean off with a rifle ball. Unfortunately the body sank.

Friday July 23

Steamed S and SW as it became clearer. Continued to ply under sail all night in the same direction among very heavy ice fields. Wind coming round to the Westward.

Saturday July 24th

Steamed SW again all day. Went through some ice that would have made Sir George Nares and the whole Arctic Committee[171] turn up the whites of their eyes. Looking back, it seemed solid as far as the horizon, and you could hardly conceive that two ships had wormed their way through it. We have one or two fainthearted ones aboard who have the terrors; it is not the going in it as the going out again, they say, and we only have a fortnight's provisions left. If we

---

168. A night of revel, along the opening lines of David McEwen Osbourne's poem of that title: "The lamps are lit, and the fire is leaping, / The fish and flesh to the board invite, / And you and I, while the town is sleeping, / Shall sup together and wake the night. / Then fill your cups ! At the dawn we're parted, / Heroes drunken with words and wine."

169. A reference to the prolific poet Bryan Waller Procter (1787–1874, also known as "Barry Cornwall"), whose songs "The Sea" and "The Stormy Petrel," with their references to whales, would have been perfect for Nox Ambrosiana aboard the *Hope*. Procter was the uncle of Conan Doyle's mentor in Edinburgh, Dr. Bryan Charles Waller, who in 1893 dedicated his own volume of verse *Perseus with the Hesperides* to the memory of his uncle and his cousin Adelaide Anne Procter, who had died aged 38 in 1864. Bryan Waller Procter had been, if not Britain's most profound poet, one of its best known. His wife was a celebrated London hostess: "beloved and admired by all who knew her," said Charles Dudley Warren's *Library of the World's Best Literature* (1896); "her house was the most popular rendezvous for literary London."

170. A Royal Navy training ship with some 300 crew and midshipmen aboard that was never seen again after sailing from Bermuda for Falmouth, England, on January 31, 1880.

171. Captain (later Vice Admiral Sir) George Strong Nares, RN (1831–1915), an explorer of both the Arctic and Antarctic. He was a junior officer on the 1852–54 attempt to find the lost Captain Franklin and his crew, in 1872–75 commanded the *Challenger* scientific voyage in the Atlantic and Indian Oceans that crossed the Antarctic Circle for the first time, and returned to the Arctic in the celebrated 1875–76 expedition, writing an account *Narrative of a Voyage to the Polar Sea*, 1878. Made a Fellow of the Royal Society, he served on the Admiralty's Arctic Committee.

got beset we should certainly have to go on uncommon short commons. We are leaving 200 miles of heavy ice between us and the sea.

Sunday July 25

A very clear day with occasional fogs. Steamed 40 miles West. Made sail in the evening. Saw a great number of 'boatswains.' We have been exulting rather during the last few days as we have been getting in to Westward very well, but our way seems to be barred now by an immense chain of flaws, which we hope to circumvent.

Monday July 26

Sailed West and South West. Made our longitude 6¼° w, and our latitude 73.56 N. Captain went aboard *Eclipse* in evening. Water swarming with food, but no animal life to be seen save 9 maulies and a school of Phoca Vitulina.[172] Wrote a POM, about a Meerschaum Pipe

> It lies within its leather case
>   As it has lain in years gone bye,
> Trusty friends and comrades true,
>   Are that old meerschaum pipe and I.
>
> For it was young when I was young
>   And many a jovial reckless night,
> We students drank, and smoked and sang,
>   While yet my meerschaum pipe was ~~young~~ white.
>
> And it was hardly brown before,
>   From home and friends I first did part,
> But bound for Russia's hostile shore,
>   I bore my meerschaum next my heart.
>
> And there upon the bloodstained ground,
>   Where many came and few went back,
> With death and pestilence around
>   Twas there I smoked my meerschaum black.
>
> And when the day our Colonel died,
>   We charged and took the Malakoff,[173]
> A Russian bullet grazed my side,
>   And shot my meerschaum's amber off.

---

172. Harbour, or common seals.
173. The Battle of the Malakoff Heights, September 7, 1855, a victory over the Russians that ended the siege of Sevastopol in the Crimean War.

But I am grizzled now and bent,
   Death's sickles near – His crop is ripe,
I fear him not but wait content,
   I wait and smoke my meerschaum pipe

ACD.

Tuesday July 27

Plying under sail about SSW. Latitude at noon gave us 73.29 N. A large finner whale, the first we have seen for some time, came up below the quarter boats. It seems to be a disputed point whether they are a good or a bad sign, the majority affect the latter opinion, but Captain David Gray throws his very weighty verdict on the minority. From my own experience I should say that the presence of finners is not by any means a bad sign.

[DRAWING 'Flinching a fish sketched by Capt. David/SS *Eclipse*']

Blew a fresh breeze in the evening, ice moving at a great rate. Spent some time in the halfdeck. *Erik* built a house as a depot in Davis Straits. On returning one season they found a polar bear lying asleep in one of the beds on the top of the blankets. Reading Maury's "Physical Geography of the Sea."[174] He explains the weed of the Sargasso Sea (in the triangle between Cape de Verdes, Azores and Canaries) by saying it is the centre of the whirl of the Gulf Stream, as when you whirl the water in a basin, you find floating corks at centre. He also remarks that railway trains always run off the line to the right hand side whether going North or South.

Wednesday July 28

Another disagreeable day. Blowing hard from the South East, which is about the worst possible direction. This is the longest interval we have ever had. The ship has not drawn blood since July 8th, except a flaw rat I shot. Blew from Eastward in the evening. As thick as pea soup and ice closing upon us rapidly. We have

---

174. By Matthew F. Maury (1806–73), an American naval officer and maritime scientist who was the first Superintendent of the U.S. Naval Observatory, but fought for the Confederacy. He finished his career at Virginia Military Institute, and his honours include three U.S. naval vessels named for him. *The Physical Geography of the Sea*, 1855, a major work of its time with a chapter addressing "The Open Sea in the Arctic Ocean," and looking at the possibilities of reaching the North Pole someday.

Conan Doyle may have seen himself in the following from Maury's introduction: "A very clever English shipmaster [Robert Methren, *Log of a Merchant Officer*, 1854], speaking recently of the advantages of educational influences among those who intend to follow the sea, remarks: 'To the cultivated lad there is a new world spread out when he enters on his first voyage. As his education has fitted, so will he perceive, year by year, that his profession makes him acquainted with things new and instructive. His intelligence will enable him to appreciate the contrasts of each country in its general aspect, manners, and productions, and in modes of navigation, adapted to the character of coast, climate, and rivers. He will dwell with interest on the phases of the ocean, the storm, the calm, and the breeze, and will look for traces of the laws which regulate them. All this will induce a serious earnestness in his work, and teach him to view lightly those irksome and often offensive duties incident to the beginner.'"

hopes that there is the open sea to the South of us from the fact that seals are coming through from the South. I thought too there was a swell from the same direction, which would settle the question.

[DRAWING 'A Right and Left among the Loons']

A very anxious and disagreeable night for us all, blowing hard, thick fog and ice everywhere. Captain and I could not turn in till 4 AM.

Thursday July 29

Horrible contemptible pusillanimous thickness over all. Made fast to a flaw, and waited for better days. Went on a journey over the ice, accompanied by our Newfoundland Sampson. Were out of sight of the ships and had great fun. Came across a most extraordinary natural snow house, about 12 feet high, shaped like a beehive with a door and a fine room inside in which I sat. Traveled a considerable distance, and would have gone to the Pole, but my matches ran short and I couldn't get a smoke. Got a long shot at a boatswain but missed him. Steamed SE when it cleared, but as it grew thick again we had to anchor once more. *Eclipse* shot at a bladder but missed it. Got a curious fungus on the ice. Gin and tobacco at night.

[DRAWING 'Natural Ice house. Lat 73.15. Long 6 w.']

Friday July 30

Suffered for the gin and tobacco. A most lovely day 72.52 N. Jan Mayen bearing SW about 100 miles and not visible. Steaming SSE at 6 knots. Took no dinner but went to the masthead in preference, enjoying a pipe and the welcome sunshine. Fell in with one or two small bladdernose seals of which we shot two, one fell to my rifle, the other was the object of the worst exhibition of shooting I ever had the misfortune to witness. I fired my only cartridge at a long range and missed, whereupon two harpooners took the job in hand, and fired 3 shots each, or 7 shots in all before the unfortunate seal dropped its head.

Saturday July 31st

Out in the open sea pitching and tossing like Billy and with her head WSW bound for the Bottlenose Bank. It is very problematical whether we will get any of the creatures, as I suppose they shift their ground like all other animals in these regions, and because Captain David got them there in April is no reason why we should see them again in August. No ice in sight. I shall never again see the great Greenland floes, never again see the land where I have smoked so many pensive pipes, where I have pursued the wily cetacean, and shot the malignant bladdernose. Who says thou art cold and inhospitable, my poor icefields? I have known you in calm and in storm and I say you are genial and kindly. There is a quaint grim humour in your bobbing bergs with their fantastic shapes. Your floes are virgin and pure even when engaged in the unsolicited 'Nip.' Yes, thou art virgin, and drawest but too often the modest veil of Fog over thy charms.

I can apostrophize the icefields, but hang the word will I say in favour of Spitzbergen, the Jotunheim of the Scandinavian mythology[175] which I saw in a gale and left in a gale, a barren rugged upheaval of a place. Sailed West and Sou'West all day. It fell calm in the evening and we lay in a long rolling swell, our sails flapping and a thick mist around.

Eclipse out of sight – has probably been steaming in the fog all night. Steamed W and SW through calm water and thick mist. We hope we may find bottlenose whales about 80 miles SE of Jan Mayen, and from there to Langaness in Iceland.[176] Keeping up our spirits. Saw some driftwood today. Hove a bottle overboard in the evening with our longitude and latitude and a request to publish where it was picked up. Bottlenose fishing has never as yet been at all developed, several ships have tried it in a half hearted way and failed. The *Jan Mayen* got 9 in 6 weeks which did not pay them, Captain David this year got 32 in a month which did pay him. Fell in with very greasy water tonight, with a strong smell of herrings and swarming with clios, I caught about 100 of the shelled variety. One would think the bottlenoses would be near such tempting grub. Heard a finner whale blowing away in the mist like an empty beer barrel. Lat 70.59. Long 0° 15 E. Passed 2 dead maulies and another bit of driftwood from Siberia. Several more finners seen.

[NO AUGUST IST ENTRY]

Monday August 2nd

Sea calm and hardly any wind. The top of Mount Beerenberg[177] is in sight, bearing WNW about 80 miles Saw several puffins, sea swallows and eider ducks, birds only seen in the vicinity of land. About two o'clock four bottlenose whales, two old and two young, came in sight and two boats were lowered away in

---

175. In Norse mythology, Jötunheim is home to frost giants and stone giants who threaten mortals and gods alike. "Spitsbergen, with its black crags and white glaciers," remembered Conan Doyle in "Life on a Greenland Whaler," is "a dreadful looking place … I saw it myself for the first and last time in a sudden rift in the drifting wrack of a furious gale, and for me it stands as the very emblem of stern grandeur."
176. A thinly populated peninsula on Iceland's northeast coast famous for its seabird populations.
177. The world's northernmost volcano, on Jan Mayen Island.

pursuit. They made straight for Cane's boat but when within shot they dived, and though we pursued them two hours we never got another chance. About 5 o'clock two more came up and Colin was sent after them but they disappeared. The *Eclipse* is in sight and had his boats' away also without success. They are funny looking brutes in the water, with high dorsal fins like finner whales. They are worth about £60 each. Quite warm now, have all our flannels off.

[DRAWING "Bottlenose whale in water"]

Tuesday August 3d

Things don't look as well this morning as there is more wind and not so many birds or food in the water. Sailing Westward. Nothing seen during the day.

Wednesday August 4th

Came into better ground this morning, there being very many birds and much grease on the water. Watched the Bosun gulls, who are very bad fishers, chasing the poor old kittiwakes until they disgorged their last meal, which the bullies devour in its semidigested condition. Sea was swarming with cetaceans about noon which we lowered away 2 boats for thinking they were bottlenoses, but they proved to be young finner whales, worthless brutes and so powerful that they would run out all our lines, so the boats were recalled. Captain shot a "boatswain." Saw many Eider ducks. Several swordfish also seen. One of them was chasing a finner whale round the *Eclipse*. The poor brute was springing right out of the water and making an awful bobbery. Carner put a rifle bullet into one young one about 40 feet long, which went away in a great hurry to tell its ma what they had been doing to it. This sea from Jan Mayen to Iceland might be called the Feather Sea. The surface is literally covered with feathers in many parts. The bottlenosing is an awful spree.

[DRAWING 'Hope in a calm among cetaceans. Aug 4th 1880.']

Was called up about 11 PM by the Captain to see a marvelous sight. Never hope to see anything like it again. The sea was simply alive with great hunchback whales, rather a rare variety, you could have thrown a biscuit onto 200 of them, and as far as you could see there was nothing but spoutings and great tails in the air. Some were blowing under the bowsprit, sending the water on to the fore-castle, and exciting our Newfoundland tremendously. They are 60–80 feet long, and have extraordinary heads with a hanging pouch like a toad's from their under jaw. They yield about 3 tons of very inferior oil, and are hard to capture, so that they are not worth pursuing. We lowered away a boat and fired an old loose harpoon into one which went away with a great splash. They differ from finner whales in being white under fins and tail. Some of them gave a peculiar whistle when they blew, which you could hear a couple of miles off.[178]

[DRAWING 'School of Hunchback whales south of Jan Mayen.']

Thursday August 5th

Nothing seen today. A stiff breeze arose towards evening and pitched and tossed us about confoundedly. We think the *Eclipse* has gone home. Steering SW.

Friday August 6th

Gave it up as a bad job and turned our head ESE for Shetland. Dense fog and rain with very little wind. Utterly beastly weather. We are all dejected at having to turn home with so scanty a cargo, but what can we do? We've ransacked the country and taken all we could get, but this is an exceptionally unfavourable year owing to the severity of last winter which has extended the Greenland ice far to the Eastward, and locked the fishes' feeding ground inside an impenetrable barrier. Here is our whole game bag for the season according to my reckoning

2 Greenland whales
2400 young seals
1200 old seals
5 polar bears
2 Narwhals
12 Bladdernoses
3 Flaw rats
1 Iceland Falcon
2 Ground Seals
2 King Eider ducks
2 Eider ducks
1 Boatswain
7 Roaches
23 Loons
1 Burgomaster
8 Snowbirds
3 Kittiwakes.

---

178. In the official log of the *Hope*, Conan Doyle gave this only the following lines: "Fell in with an extraordinary number of humpback whales. Hundreds of them were feeding and blowing round the ship and even under the very bowsprit. Useless for commerce." (Lubbock's *Arctic Whalers*, p. 409.) Judging from his personal log, by contrast, there was a writer in the making.

[DRAWING 'Sampson and the Hunchback whale']

Saturday August 7th

Groping homeward under steam and sail in such a thick fog that we can hardly see the water from the side of the ship. Took in the two funnel boats. We have not got our reckoning now for several days, and as we have been dodging about zigzag after these bottlenoses, our dead reckoning is very uncertain. It isn't nice to be steaming along in the North sea in a fog with Iceland and the Faeroe Islands knocking about in front of us. Several puffins and other land birds seen.

Sunday August 8th

Cleared up a little although it was raining nearly all day. Had a mackerel line over all evening but got nothing. Sighted land about 8 PM which proved to be the north end of Faeroe island. A nice job if we had come on it in the dark. Saw a schooner running North about midnight, probably bound for Iceland from Denmark. Men busy drying our whale lines.

Monday August 9th

A beautiful clear day with a blue sky and a bright sun. Wind from the NE, a good strong breeze before which we are flying homeward with all sail set, and the bright green waves hissing and foaming from her bows. No mackerel again. Ship all covered with whale lines drying. Expect to make the land late tonight. Saw a Solan Goose and a little bird called a Stienchuck, also some stormy petrels. The kittiwakes down here are a smaller breed, I think, than those further north. All hands on the lookout for land.

Sunday August 10th

Up at 8 AM to see the land bearing WSW on the starboard bow. Half a gale blowing and the old *Hope* steaming away into a head sea like Billy. Hence the feebleness of my handwriting. The green grass on shore looks very cool and refreshing to me after nearly 6 months never seeing it, but the houses look revolting. I hate the vulgar hum of men and would like to be back at the floes again.

> "There is society where none intrudes
> Upon the sea, and music in its roar!"[179]

---

179. "There is society, where none intrudes, / By the deep sea, and music in its roar!" from Byron's *Childe Harold*, published 1812–18. But in a diary Conan Doyle kept on another voyage as ship's doctor, to West Africa, he quoted its next verse in a different frame of mind: "Byron said 'Roll on thou deep and dark [ocean]' and the deep and dark is doing it

Passed the skerry light, and came down to Lerwick but did not get into the harbour as we are in a hurry to catch the tide at Peterhead, so there goes all my letters, papers and everything else. A girl was seen at the lighthouse waving a handkerchief, and all hands were called to look at her. The first woman we have seen for half a year. Our Shetland crew were landed in four of our boats and gave 3 cheers for the old ship as they pushed off, which were returned by the men left. Lighthouse keeper came off with last week's weekly *Scotsman* by which we learn of the defeat in Afghanistan. Terrible news.[180] Also, that the *Victor* has 150 tons the dirty skunk.[181] Took our boats aboard and went off for Peterhead full pelt. Fitful head, and Sumburgh light[182] twinkling away astern like a star. Herring fishing seems to be a success. Saw a large grampus.

Wednesday August 11th

Dead calm and the sun awfully awful. Saw Rattray head[183] at 4 PM. The sea black with fishing boats. Hurrah for home! Pilot boat came off at 6 PM and we lay off for high water at 4 in the morning. Hundreds and hundreds of herring boats around us.[184] Crew getting on their shore togs. Well, here we are at the end of the log of the *Hope*, which has been kept through calm and through storm, through failure and success; every day I have religiously jotted down my

with a vengeance. I have no reverence for the ocean, and it seems to have none for me. If you think of it, it is only an endless repetition of two molecules of Hydrogen to one of Oxygen, with salts in suspension."

180. "A terrible and most unlooked-for disaster has befallen the British arms in Afghanistan," began the *Scotsman* account of July 29, 1880, headed "Disaster in Afghanistan / Severe Defeat of Burrows' Brigade / Retreat on Kandahar." A British force of some three thousand had been close to annihilated at Maiwand by Pathan tribesmen. It made a lasting impression on Conan Doyle. Six years later he started writing a tale called *A Study in Scarlet*, set in London in 1881, and made his narrator a former army surgeon, Dr. John H. Watson:

"I was duly attached to the Fifth Northumberland Fusiliers as Assistant Surgeon. The regiment was stationed in India at the time, and before I could join it, the second Afghan war had broken out. On landing at Bombay, I learned that my corps had advanced through the passes, and was already deep in the enemy's country. I followed, however, with many other officers who were in the same situation as myself ....

"The campaign brought honours and promotion to many, but for me it had nothing but misfortune and disaster. I was removed from my brigade and attached to the Berkshires, with whom I served at the fatal battle of Maiwand. There I was struck on the shoulder by a Jezail bullet, which shattered the bone and grazed the subclavian artery. I should have fallen into the hands of the murderous Ghazis had it not been for the devotion and courage shown by Murray, my orderly, who threw me across a pack-horse, and succeeded in bringing me safely to the British lines."

Now home, uncertain about his future, and looking to share the expense of lodgings, Watson is introduced to someone described as "a little queer in his ideas – an enthusiast in some branches of science."

"How are you?" says the man he meets in the laboratory of St. Bartholomew's Hospital: "You have been in Afghanistan, I perceive." It is Sherlock Holmes, beginning the partnership that would bring A. Conan Doyle literary fame and fortune.

181. The following year the *Hope* fared far better than the *Victor*. According to Lubbock's *Arctic Whalers*, p. 410, in 1881 it took nine whales, four bottlenoses, five thousand seals, thirteen polar bears, and two Arctic foxes, while the *Victor*, reports Buchan's *Peterhead Whaling Trade*, p. 32, was lost in the Arctic ice.

182. An 1821 lighthouse, the oldest in the Shetland Islands, on the southern tip of the main island.

183. A headland on the Aberdeenshire coast, marking the *Hope*'s return to Scotland proper from its voyage. Peterhead was now less than ten miles away.

184. In fact a record year for another important Peterhead industry, with over 700 boats taking part during the season, and ashore at some fifty curing sites, between mid-July and the first week of September. The herring taken that summer were more than twice the previous year's.

impressions and anything that struck me as curious, and have tried to draw what I have seen. So here's an

End of the log of the SS Hope.

Our Illustrations

Fresh Meat[185]
Freemason's Flag
Ramna Stacks
A Peterhead Whaler
Sealing Costume
The Hope among loose ice
A family of Seals
Seal Knife
Bear's Footmark
Ships among the Seals
Seal Club
Sketches at Young Sealing
Milne's Funeral
Our Hawk
Saturday's night at sea
A Snap Shot
Our Evening Exercise
All Hands over the bows
Plan of Seal fishing
Effects of refraction
Five Bulls at 100 Yards

Vol II

Old Sealing
Making off Seal's blubber
Not sold but given away
Poking the mob boat
Harpoon Gun
Hope off Spitzbergen in a gale

---

185. These drawings, and some others unlisted by Conan Doyle here, will be found in his log, but not the very first one ("Fresh Meat"). The next one ("Freemason's Flag") appears in his ninth entry dated March 7th. It does not seem that "Fresh Meat" could have illustrated any of the previous entries, given the manuscript's appearance, and the ordering of the list suggests that it appeared too early in the account to depict a whale or seal hunt. Perhaps it was a frontispiece of sorts – the caption suggests it could have been ruefully self-descriptive – that he later carefully removed from the manuscript for some reason.

The Bear we did <u>not</u> shoot
Greenland Sword fish
The Lesser Auk (loon)
Swordfish chasing Seals
A Capture
John Thomas & his friend
Towing the Narwhal home
A Narwhal
Maulie stealing our Roach

Vol III

Capture by Eclipse's boats
Our bear
Bear and Shark devouring Narwhal
Boats after 2 fish
Our First Fish
Whale and 2 fast boats
Buchan's Boat
Ship flinching a whale (Capt J Gray)
  [ditto]   (Capt David Gray)
A Right and Left among Loons
Natural Ice House
Bottle Nose Whale
Hope among Cetaceans
Hunchback Whales
Sampson and the Fish

Game Bag of the Hope (continued)

July	1	1 Narwhal
	2	
	3	
	4	
	5	
	6	1 Flaw Rat 7 Loons 1 Roach 1 Kittiwake 2 Snowbirds
	7	
	8	a Greenland Whale
	9	
	10	

```
11
12
13
14
15
16
17
18
19
20
21
22
23
24
25
26
27
28
29
30
31        2 Bladdernose seals
August 1
       2
       3
       4    a Boatswain
       5    2 Eider ducks
```

My Own Gamebag

Young Seals and Young Bladders xxxxxxxxx. xxxxxxxxxxx xxxxx
                    xxxxx. xxxxxxxx
Old Seals xxxxxxxxxx. xxxxxxxxxx. xxxxxxxxx
Bladdernoses xxx
Loons xxxxxxxxxx.
Roaches xxx
Maulies x
Snowbirds xx
Kittiwakes xxxx
Flaw Rats xx

Courtesy of the Surgeons' Hall Museum, Edinburgh.

Licensed to kill

Courtesy of Conan Doyle Estate Ltd.

"It was quite an ovation"
Conan Doyle revisits the Arctic

Our six-month being done, we tie up again,
   And the lads all go ashore,
With plenty of brass and a bonny bonny lass
   For to make them taverns roar!

To Greenland's coast we'll drink a toast,
   And to them we hold most dear;
And across the icy main to the whaling grounds again
   We'll take a trip next year!
        "Greenland Bound" (traditional chanty)

CONAN DOYLE had taken to the life and work so well that Captain Gray offered him double pay as harpooner as well as surgeon if he would return for the 1881 season. "Never had such a jolly time in my life," Conan Doyle said about the *Hope*, but he kept to Edinburgh instead, receiving in May 1881 his first medical degree, the Bachelor of Medicine and Master of Surgery. A full-fledged M.D. required a written dissertation that would come in 1885, but an M.B.C.M. entitled him to practise medicine. He posed in cap and gown for a dignified graduation photograph, but also drew a gleeful sketch of himself brandishing his prized diploma with the caption "Licensed to kill." Now 22, he went forth into the world without a clear idea of exactly what he would do, except to start earning his own livelihood and contributing to his family's welfare.

"When a man is in the very early twenties he will not be taken seriously as a practitioner," Conan Doyle would learn during the next few years, "and though I looked old for my age, it was clear that I had to fill in my time in some other way." Over the months following graduation he considered a number of career paths, including service in the Royal Navy as a medical officer. He also undertook another voyage as ship's surgeon, aboard the passenger-carrying freighter S.S. *Mayumba* from

Liverpool to West Africa and back, October 1881 to January 1882. "I could in this way see something of the world," he wrote, "and at the same time earn a little of the money which I so badly needed if I were ever to start in practice for myself." But at its end, he felt disinclined to continue this type of work. "I have turned up all safe," he reported in a crisp note home, "after having had the African fever, been nearly eaten by a shark, and as a finale the *Mayumba* going on fire between Madeira and England, so that at one time it looked like taking to our boats and making for Lisbon."

In time he hoped to win a surgeonship in a big city hospital, preferably in London, but competition for such posts was intense. He decided to cast his lot in the meantime with his remarkable and not overly scrupled fellow student from Edinburgh, Dr. George Budd (see p. 243 n110). In Plymouth, Budd boasted, he had created a practice worth a splendid £3,000 a year; and what Budd claimed was true, Conan Doyle found when he joined him as his assistant – but at the expense of medical ethics. The association lasted just six weeks, ending by chicanery on Budd's part, with Conan Doyle departing to start a practice of his own somewhere else.

Conan Doyle was young, inexperienced, and knew nobody in Portsmouth, but it was the largest city on England's Channel coast. He took a steamer there in June 1882, and rented a house in its residential district of Southsea to serve as both home and surgery. "I move into my house tomorrow, No. 1 Bush Villas, Elm Grove," he told his mother: "I have a few shillings left to live on and have put £5 by for the rent." To a family friend in Edinburgh, Mrs. Charlotte Drummond, he summed up his situation in humorous terms:

> I took the most central house I could find, determined to make a spoon or spoil a horn, and got three pounds worth of furniture for the Consulting Room, a bed, a tin of corned beef and two enormous brass plates with my name on it. I then sat on the bed and ate the corned beef for a period of six days at the end of which time a vaccination turned up. I had to pay 2/6 for the vaccine in London, and could only screw 1/6 out of the woman, so that I came to the conclusion that if I got many more patients I would have to sell the furniture.... I write away for the papers in the intervals of brushing floors, blacking boots and the rest of my labours – occasionally glaring out through the Venetian blinds to see if anyone is reading the plate.

He kept up appearances clandestinely, at a time when barely able to keep himself fed: "I have to sit up nearly to midnight every night in order to polish my two doorplates without being seen," he told his mother. "Have no gas yet – but candles." He was determined to succeed, though, and endeavoured to supplement his meagre income from medicine with whatever he could also earn from writing.

Conan Doyle outside his house, 1 Bush Villas, Southsea, in the mid-1880s. (At the windows, household staff and his younger brother, Innes.) Conan Doyle lived and practised medicine here from June 1882 until the end of 1890, and wrote the first Sherlock Holmes tale here in 1886. (Courtesy of Conan Doyle Estate Ltd.)

Bush Villa — Southsea

Before long came several shipments from home of books and personal possessions which he used to decorate his consulting room. "Have just opened the box and never was more astonished in my life," he told his mother. "You would be astonished if you could see my C.R. now. It looks awfully well. You have done splendidly. The Arctic things were especially useful." These were souvenirs from the *Hope*, listed in his log for July 1–2, 1880. It's likely he was the only doctor in the south of England displaying a bladdernose seal's bones in his consulting room, but the Arctic theme served him well as he worked away at fiction while waiting for patients, for while his initial efforts were met with discouragement – "I have got a lot of literary chickens hatched and flying about, but none of them have come home to roost yet" – it turned out that his experience at sea was the key to his first successes.

The first of these, in the second half of 1882, was a psychological ghost story called "The Captain of the Pole-Star," set in the Arctic on a haunted ship. For a youngster his age it was a remarkable feat. To Mrs. Drummond, he wrote in

January 1883: "Literature has been good to me too of late. I should like your opinion on 'The Captain of the Pole-Star' in January 'Temple Bar' – a ghost story. They sent me ten guineas & a copy, so I think they liked it." Its success encouraged him to address the Arctic in non-fiction also. "Am writing a leader on Sealing for the *Daily Telegraph* – will write another on Whaling if successful," he told his mother about this time. That appears to have fallen through, but he returned to the sea in fiction again, this time drawing more on his second experience at sea, out to West Africa. "I have sent my 'Statement of J. Habakuk Jephson M.D.' to Cornhill – May luck go with it!" he wrote to his mother on June 15, 1883, a year after his arrival in Portsmouth. In due course that story, inspired by the *Mary Celeste* found mysteriously abandoned in the eastern Atlantic in 1872, was accepted by *The Cornhill*, Britain's foremost literary magazine, edited by James Payn, an idol of Conan Doyle's.

Its acceptance led to his introduction to London literary society at a full-dress dinner for *Cornhill* contributors that autumn, at a famous Greenwich tavern called, appropriately enough, The Ship. "I remember the reverence with which I approached James Payn, who was to me the warden of the sacred gate," said Conan Doyle in *Memories and Adventures* – and "I came back walking on air." So it sounded also from his report home immediately after: "Everybody was very charming and we all got along most famously. Everyone seemed to be a great & shining light except poor me." It whetted his appetite for a permanent place among such men, he soon reporting that: "I am going to read a lecture this winter before the literary & scientific society – I think on the American Humourists but have not quite made up my mind."

Portsmouth provided many opportunities during the eight-and-a-half years Conan Doyle lived there, but its Literary & Scientific Society would be of special value to him, making him known to the town's intelligentsia and leading to wider intellectual horizons, new friends from different walks of life, and more patients for his practice. One new friend was the Society's chairman, retired Major General Alfred Drayson, an amateur astronomer whom Conan Doyle compared favorably to Copernicus. Drayson also encouraged the younger man's budding interest in psychic research, which would become the principal mission of his later years. "I have many pleasant and some comic reminiscences of this Society," Conan Doyle said in *Memories and Adventures*:

> We kept the sacred flame burning in the old city with our weekly papers and dis-cussions during the long winters. It was there I learned to face an audience, which proved to be of the first importance for my life's work. I was naturally of a very

nervous, backward, self-distrustful disposition in such things and I have been told that the signal that I was about to join in the discussion was that the whole long bench on which I sat, with everyone on it, used to shake with my emotion. But once up I learned to speak out, to conceal my trepidations, and to choose my phrases.

But he decided not to debut on American Humourists on December 4, 1883. He chose instead a topic based on personal experience, adding in *Memories and Adventures* that over time:

> I gave three papers, one on the Arctic seas, one on Carlyle and one on Gibbon. The former gave me a quite unmerited reputation as a sportsman, for I borrowed from a local taxidermist every bird and beast he possessed which could conceivably find its way into the Arctic Circle. These I piled upon the lecture table, and the audience, concluding that I had shot them all, looked upon me with great respect. Next morning they were back with the taxidermist once more.

"It is to be quite a swell affair," he told Mrs. Drummond, and she gave him long-distance assistance with his apparel for the occasion, to make a suitable impression upon his audience. He was delighted with the result. "What a villain I am not to write by return and tell you how pleased I was with the shirt," he wrote her a day or two before the talk:

> but you know I am always an erratic correspondent and indeed I have been over ears in work for the last week or so. The collar too is a masterpiece. I have a crutch stick of ebony and silver which I won as a prize and with the collar I am more than a masher – I am a dude – which is an Americanism for the masherest of mortals.
>
> The lecture is ready now … I do hope it will pass off well. It will do me a lot of good in the way of getting my name known among nice people. I have a sturdy phalanx of bachelor friends, strong armed and heavy sticked who may be relied upon for applause.

The main thrust of his "Arctic Seas" talk was the quest to reach the North Pole, and what he believed to be the right way to do it. In the Portsmouth Literary & Scientific Society minute-book for 1883, at Portsmouth Central Library now, are three newspaper accounts of his talk at Penny Street Lecture Hall that 4th of December, the most comprehensive of which was the *Hampshire Telegraph*'s on the 8th:

> ### "The Arctic Seas"
>
> Dr. A. Conan Doyle read a paper on the above subject. At the outset the lecturer remarked that great as had been the strides which our knowledge of the world had made during the last century, there was still ample employment for the traveller and the geographer of the future. Round the Pole 2½ million square miles had been untouched by the foot of man, whilst away down in the Antarctic regions the great mysterious continent of the South shrouded itself behind a veil of ice. From

the highlands of Thibet to the stony plains of the interior of Australia, and from the great lakes of Africa to the savannahs of Central America, there were blank spaces on the map which were an opprobrium to science and a challenge to human daring. Gradually, as the years rolled by, the limits of those *terrae incognitae* became somewhat contracted, whether by the slow march of commerce and civilisation, or the meteor-like passage of some daring passenger. The process, however, was slow, and they would hand down to their descendants a legacy of the unsolved riddles of nature almost as rich as we had inherited, which with increased appliances and modes of progression they might succeed in solving where their fathers had failed. In all the long and thrilling annals of travel and discovery there was nothing which could equal in dramatic interest the struggle made by the human race to reach the North Pole, and the desperate efforts during the past two hundred years must strike the imagination of the most thoughtless. It was not a pleasing story. It was a record of blasted hopes and baffled exertions; of crushed ships, starvation, scurvy, of privation, and too often of lonely graves far up in the dim twilight land. But there was a brighter side to the question, because was there not also a record of indomitable pluck, wonderful self-abnegation and devotion – a training-school for all that was high and godlike in man? The spectacle of a long succession of men who had crowded forward anxious to sacrifice their own individuality for the common good, risking their lives with a light heart in the interests of science, was surely one which pointed to something higher in human nature than pessimists would have them believe. Having given a description of the Arctic seas, much being from personal experience, Dr. Doyle proceeded to enter into the voyages made by the old Elizabethan captains, tracing in their order the various efforts of Davis, Baffin, Hudson, Parry, and others in their endeavours to reach the Pole, and adding additional interest by detailing the most salient points of these expeditions. Coming to the British expedition of 1875, he remarked that the present generation of English seamen had pushed their ships not only beyond those of their hardy Transatlantic and Continental rivals, but beyond the utmost limits reached by their forefathers, thus showing that in what some were pleased to call "these degenerate days" our seamen had shown themselves to be of the same mettle as of old, and even the most inveterate *laudator temporis acti* could hardly ignore such stubborn facts as degrees of latitude. Civilised man had of late years been within 399½ miles of the Pole. Would he ever reach it? He saw no reason to doubt it; the question hinged on whether the remaining 390 miles were as bleak and as barren as those which had been traversed. He was inclined to hold the opinion that they were not, and that after a certain point the temperature should change for the better as they approached the Pole. He knew the opinion was scouted by many great Arctic authorities, yet at the same time it was held by some of the whaling captains who had spent their lives in those seas. Amongst his reasons for so thinking was the statement of Morton,[1] who actually saw the open water, and as that sea would be bounded by the ice, it would vary with the intensity of the season, sometimes being much encroached and sometimes expanded. Then ice had a uniformly southern drift, so there must be a

---

1. William Morton who served in two American expeditions, 1850 and 1853, financed by Henry Grinnell, to rescue Captain Sir John Franklin, publishing *Dr. Kane's Arctic Voyage: Explanatory of a pictorial illustration of the second Grinnell expedition* in 1857. (The second Grinnell expedition was led by Dr. Elisha Kent Kane.)

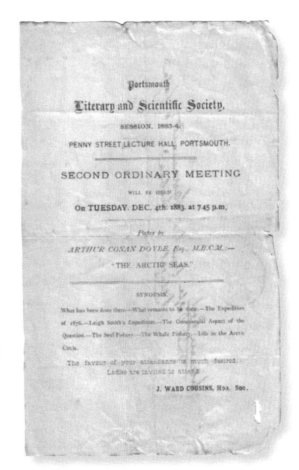

"It is to be quite a swell affair," and the synopsis given on the notice for the talk indicates how Conan Doyle had organized his thoughts: "What has been done there – What remains to be done – The Expedition of 1876 – Leigh Smith's Expedition – The commercial aspect of the question – The seal fishery – The whale fishery – Life in the Arctic Circle." Shown is Conan Doyle's own copy of the notice. (Courtesy of Conan Doyle Estate Ltd.)

clear space in the centre, and as the earth was not perfectly round, but flattened at the poles, those therefore were nearer the centre of the earth by an appreciable difference than we were. The nearer the centre of the earth was approached – in coal mines or otherwise – the higher was the temperature, consequently one would think that the immediate circumpolar region would receive an additional supply of heat from the centre of the earth, which might counteract the lack of heat from above. Having described whale catching and other interesting items in connection with the Arctic seas, Dr. Doyle asked whether the initial idea had been correct, and whether the thing had been attempted in the manner which was most calculated to lead to success? That was open to doubt, but it was the fact that all the ice in the Arctic seas drifted southward, and if they examined the geography of those seas the very first thing that struck them was that Smith's Sound and Kennedy Channel was the very worst highway possible to a northern latitude. Why should that approach to the Pole be chosen in preference to the broad room between Greenland and Spitzbergen, where there was a fine expanse of sea, in which the floes had plenty of room to circulate? If the winter, however, were a hard one and the summer mild there would, he granted, be little chance of success, but if the reverse were the case, he thought it certainly a better highway. If the Government wished to reach the Pole they should go to no expense in elaborate preparations, but commission one of the many suitable cruisers belonging to the nation and make use of a few of the gallant officers on half-pay. Let the vessel perform the voyage every year, starting about the middle of June, and on reaching the ice barrier, steam along and look at it and not commit the error of running the ship into the first fields arrived at. They would probably come back unsuccessful in two months, but let them go next year and the next, and sooner or later there must come a warm season, with violent gales from the north, splitting up the barriers. Then was the chance, and could an opening be found two days' steaming would take them to the Pole. He thought the Government should put a reward of at least £30,000 for the discovery,[2] and though some might ask the good in reaching it, he would reply that that was a narrow minded view of the matter, as such attempts kept up the spirit and pride of a nation. Economists and utilitarians might argue as they pleased, but he knew that a glow of pride and joy would fill the hearts of the great Anglo-Saxon race when the day came (as he believed it would) when the flag they all loved should be hoisted on the northernmost pinnacle of the earth.

THE PRESIDENT[3] moved a vote of thanks to the lecturer, and said a point which had struck him very much was the idea put forward by Dr. Doyle of there being an open space up to the Pole. He had always entertained that idea himself. He did not quite agree about the heat being greater at the Poles in consequence of their being nearer the centre of the earth, because it was a moot question whether there was that increase of heat in the earth which some of the old theorists thought. He should like to know the opinion of the lecturer as to whether there had been a sufficient amount of use made of stationary balloons, as by their means he thought a very much better track might be selected than going blindly on as they now had to do. Another interesting point was as to the use of the Pole if found. Well, they could never tell the use of a discovery until it was made. (Hear, hear.) He believed it would be, especially by that practical suggestion of the lecturer of going every year. – The HONORARY SECRETARY[4] seconded the vote, and said few would credit the number of Polar expeditions which had been made, but he believed there had been between two hundred and three hundred explorations to find the North-West passage. The number of expeditions to discover the Pole had been much less than those which had commercial purposes for their object. He endorsed the observation of the Chairman that the explorations did an immense amount of good. (Hear.) All knowledge was power, and physiological science would amongst others gain advantages by such explorations and discoveries. There was one important matter which had been proved, and that was that the severe temperature of the Arctic seas could be endured without the use of alcohol (hear, hear), so that if abstainers

---

2. "The Government reward of £5000 was not tempting enough for whalers," he said, according to the *Hamphire Post*'s account of December 7th, "seeing that every whale caught was worth £1000. He thought it ought to be increased to £30,000."
3. Major General Alfred Drayson.
4. Dr. J. Ward Cousins, F.R.C.S.

Major General (Ret.) Alfred Drayson.
(Courtesy of Morgan Family History Blog.)

had no other argument than that it would be a very strong one indeed. – The Rev. Dr. Colbourne said he believed coal fields had been found in the high Arctic regions, and if so it must argue that at some period of the earth's existence there had been a higher temperature for the existence of plants from which the coal was formed. Then again, from what race did the natives of Greenland spring. Were they the remnants of an aboriginal race? – The resolution was then carried, and Dr. Doyle, in reply, said coal fields were found in Greenland and Spitzbergen, and there were evidences of a very high temperature in times gone by, but the amount of research by scientific men did not justify any very hard and fast lines being drawn on the subject. As to the natives, he believed the general consensus of scientific opinion was that many thousand years ago all Europe was inhabited by a race of Esquimaux, who were gradually pushed away, and by superior civilisation gradually overwhelmed. As to stationary balloons for making observations, he thought it a most excellent idea.[5] (Hear, hear.) – The meeting then separated.

"The lecture is over – Gott sei dank! and was an unqualified and splendid success," he wrote his mother: "far more so than I had ever dreamed of or dared to hope for. From

---

5. In fact he closed with a well-received joke, according to the *Portsmouth Times* of December 5th: "He thoroughly concurred in the use of balloons, and believed that did a man cross the North Pole in a balloon a ship would pick him up in the seas beyond. He would, however, recommend such an explorer to insure his life. (Laughter.)" This joke was aptly chosen; at the time he was supplementing his income as a medical examiner for the Gresham Life Assurance Company, even writing occasional advertising copy for it. *Arthur Conan Doyle: A Life in Letters*, pp. 199–202.

the first word to the last the audience (which was a very crowded one[6]) followed me most closely and often I could not get on for the cheering. When I finished there was tremendous applause – a vote of thanks was carried unanimously and then speaker after speaker got up to comment on 'the splendid paper' 'the most able paper' 'the beautifully written paper' which they had heard. It was quite an ovation."

The lecture marked a turning point in Conan Doyle's life in Southsea, renewing his confidence and professional ambition. Publication of "J. Habakuk Jephson's Statement" the following month proved a further leap forward. Though it lacked his by-line, in the custom of *The Cornhill*, it paid a gratifying twenty-nine guineas. The story was widely noticed, and taken by some critics as the work of Robert Louis Stevenson, while others likened it to Edgar Allan Poe's *Narrative of Arthur Gordon Pym*. "There is a good time coming down here," he assured his mother. "We have everyone's good name and respect and affection from not a few, without as far as I know an enemy, so we are bound to succeed."

In his two most notable stories to date – "Pole-Star" and "J. Habakuk Jephson" – Conan Doyle had merged his experiences at sea with the story-telling techniques of Poe. From there, Sherlock Holmes was but a matter of time. Four years later, still dividing his time between medicine and literature, he began writing a detective tale called *A Study in Scarlet*, featuring two soon-to-be immortal characters, Dr. Watson and Sherlock Holmes. But his first two novellas featuring Holmes made little headway with the reading public at the time, and Conan Doyle still had a long hard road before him both as a writer and a physician. Now married and with a young family to support, he turned next to the first of his many historical novels, *Micah Clarke* and *The White Company*.

Then in 1891 Conan Doyle resettled in London to practise in ophthalmology, after a short period of study in Vienna and Paris. While he continued to have medical ambitions, he had also continued to write fiction even while studying to qualify as a specialist. And he gave Sherlock Holmes another chance, this time in short-story form. After publication of the first two stories in the July and August issues of *The Strand Magazine* made him a sudden literary sensation, he decided to give up medicine for literature. Even with this new renown, though, he continued to turn to the Arctic. "What a climate it is in those regions!" one interviewer of the new literary star was regaled: "We don't understand it here. I don't mean its coldness – I refer to its sanitary properties. I believe, in years to come, it will be the world's sanatorium. There, thousands of miles from the smoke, where the air

---

6. 250 ladies and gentlemen, according to the late Geoffrey Stavert's book *A Study in Southsea: The Unrevealed Life of Doctor Arthur Conan Doyle* (Portsmouth: Milestone Publications, 1987), p. 45.

is the finest in the world, the invalid and weakly ones will go when all other places have failed to give them the air they want, and revive and live again under the marvelous invigorating properties of the Arctic atmosphere."

Conan Doyle used his new prominence to give the public his views. "The Glamour of the Arctic" in *The Idler*, July 1892, took up the thread of his lecture before the Portsmouth Literary & Scientific Society in suggesting a way to reach the North Pole, drawing the admiration of the Norwegian explorer Vilhjalmur Stefansson, and leading to a lifelong friendship. "Conan Doyle is not merely a sturdier Watson and a kinder Holmes. He is also a gentler Sir Nigel and a mellow blend of all the host of his nobler characters," wrote Stefansson in a May 3, 1922, tribute in *The Outlook*. "My dear Stefansson," Conan Doyle had written at the end of his spiritualist lecture tour of Australia in 1920, "How strange our tasks! You are working on reindeer and I on disembodied spirits & both are equally part of the great whole." It is unclear if Stefansson shared Conan Doyle's view of spiritualism, though he was sympathetic in his *Outlook* article; but they recognized each other as determined, even intrepid explorers of unknown regions.

Five years later he gave a personal account of "Life on a Greenland Whaler" in the *Strand Magazine* for January 1897. In it he told the story of boxing with Jack Lamb the steward, and Lamb telling the first mate Colin McLean that Conan Doyle was the best surgeon the *Hope* had had because he had given Lamb a black eye. His article, which he would rework in 1924 as a chapter for his autobiography *Memories and Adventures*, was picked up throughout the English-speaking world. Close to home, the *Peterhead Sentinel* remarked:

It will no doubt be of interest to Dr Conan Doyle to know that both Colin McLean and Jack Lamb are still alive. The former returned from the whale fishing as usual in November last, at which time he was mate of a Dundee whaler. He is as stalwart as ever, but he talks of giving up whale hunting and settling down in Peterhead, as his eyes are not so keen of vision as in 1880. He still retains kindly recollections of the boxing surgeon of the *Hope* in 1880; and he had no difficulty in spotting Dr Doyle and the other members of the group when we submitted the photo, reproduced in the *Strand Magazine*. Also his idea of the doctor's physique is as respectful as the doctor's evidently is of his, though his opinion of the surgeon's powers in walking on an Arctic ice-field is correspondingly low. About Jack Lamb, the sweet-voiced tenor and sturdy, deep-chested steward of pugilistic tendencies, Dr Doyle's surmise is quite correct. No doubt Greenland is as much a dream to him as it now is to the erstwhile surgeon of the *Hope*; for he has returned to his old trade. He is now chief baker to the Queen and travels with her wherever she goes. If ever Dr Doyle happens to visit the Deeside Highlands in autumn, or, better still, if he can spare an hour or two to run down to Windsor, where Jack Lamb and his wife and family have their home, he will be able to renew his acquaintance with his old friend. No doubt Jack's repertoire of pathetic and sentimental songs has increased, but for the sake of old times he will be only too glad to give 'Her bright smile haunts me still' and 'Wait for me at Heaven's gate, Sweet Belle Mahone.' He may even be prevailed upon to fight literally their battles over again.

Conan Doyle referred frequently in public appearances to his Arctic voyage aboard the *Hope*, and occasionally heard from people who noticed. But even when the January 23, 1900, *Daily Mail* made no mention of the Arctic in reporting an Authors Club dinner for him, as he was about to go off to the Boer War in South Africa as a volunteer army field surgeon, he still heard from someone aware of it, the Shipping Federation's James Brown at London's Royal Albert Dock:

> Though unknown to you I have heard and read a lot about you. Being an old Peterhead skipper I knew Capt. John Gray, Colin Campbell [sic] the mate, and all the rest that were in the SS *Hope* when you were Doctor of her. And been shipmates with most of them. And when I read in today's *Daily Mail* your reply to the toast of your health at the Authors Club last night, I take the liberty of writing to wish you God Speed and Safe Return."

Conan Doyle recognized him as the master of the *Windward* who in 1896 had brought Fridtjof Nansen back to Norway safely after his unsuccessful attempt to reach the North Pole. Conan Doyle preserved this letter among his papers, along with ones he received from Jack Lamb.

The Arctic experience would also find its way into Sherlock Holmes, from his log's penultimate entry referencing the Battle of Maiwand, where he would send Dr. Watson in the first very tale, to the appearance of sailors, ships, and nautical themes in many of the other stories. "No one but an acrobat or a sailor could have got up to that bell-rope, and no one but a sailor could have made the knots," Holmes typically remarks in one case. "His face was thin and brown and crafty," Watson records in another, "and his crinkled hands were half closed in a way that is distinctive of sailors." In *The Adventure of Black Peter*, a whaling captain named Peter Carey is found gruesomely murdered with a harpoon through his chest – "pinned like a beetle on a card." In describing Carey's villainous nature, Conan Doyle not only drew upon his knowledge of whaling, but gave a subtle nod to his former Peterhead shipmates and their rivalry with the men of Dundee: the odious Carey, we are told, is a Dundee man.

Even Conan Doyle's poetry returned on occasion to the adventure of his youth. A jaunty effort called "Advice to a Young Author," published in 1911, struck a seafaring note:

First begin
Taking in.
Cargo stored,
All aboard,
Think about
Giving out.
Empty ship,
Useless trip!

Seldom was this advice more appropriate than when he befriended and encouraged Frank Thomas Bullen, whose troubled youthful wanderings had included a long voyage on a New Bedford, Mass., whaler – turned into an 1898 bestseller, *The Cruise of the Cachelot*, the first of many books by Bullen about life at sea. Bullen dedicated a 1901 novel of his to Conan Doyle, who in turn, in *Through the Magic Door*, praised *Cachelot* and its author as "sperm-whale fishing, an open-sea affair, and very different from that Greenland ice groping in which I served a seven-month apprenticeship [and] handled by one of the most virile writers who has described a sailor's life." In 1914 he wrote the preface for a collection of sea chanties by Bullen, saying:

> Like yourself I have heard them many a time when I have been bending to the rhythm as we hauled up the heavy whaling boats to their davits. It is wonderful how their musical rise and fall, with the pull coming on the main note, lightened the labour. I fear in these days of steam winches that the old stamp-and-go of ten men on a rope is gone for ever. And yet your book will help to preserve it, and to those who know and can feel, there is a smack of salt spray in every line of these rude virile verses.[7]

Bullen "was the best singer of a chanty I have ever heard," Conan Doyle mused in his 1921 travel account *The Wanderings of a Spiritualist*, after describing how the sighting of thrasher and sperm whales at sea had "aroused the old whaling thrill in my heart."

As Stuart Frank, Senior Curator of the New Bedford Whaling Museum reminds us, Conan Doyle and Herman Melville are the only great writers who have described nineteenth-century whaling from first-hand experience. To the end of his life in 1930, Conan Doyle's time in the Arctic remained a fond and vivid memory for him. "There is a glamour about those circumpolar regions which must affect everyone who has penetrated to them," he had written years earlier: "My heart goes out to that old, grey-headed whaling captain who, having been left for an

---

7. In Frank T. Bullen, F.R.G.S., and W. F. Arnold, *Songs of Sea Labour* (London: Orpheus, 1914).

instant when at death's door, staggered off in his night gear, and was found by his nurses far from his house and still, as he mumbled, 'pushing to the norrard.'" Not long before his death Conan Doyle drew a whimsical sketch called "The Old Horse," showing himself as a broken-down draft animal pulling a heavy cart filled with the freight of a busy and active life – "medical practice," "Sherlock Holmes," "historical novels," and much more, and at the top of the pile, "Arctic." In the background, along a winding trail that depicted additional milestones, he included a boatful of harpooners pulling on a whale.

Conan Doyle died at home, surrounded by his family, on July 7, 1930. At last, he too was "pushing to the norrard."

Jon Lellenberg & Daniel Stashower

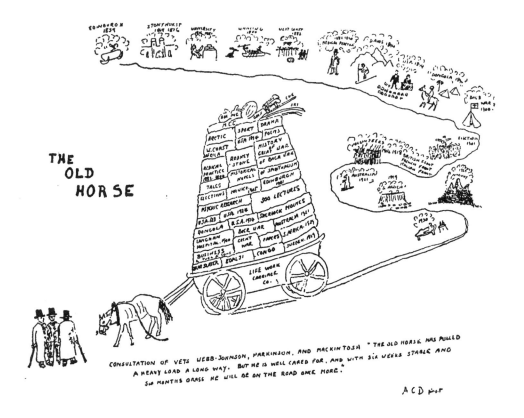

(Courtesy of Conan Doyle Estate Ltd.)

Arthur Conan Doyle's
Arctic writings

The Glamour of the Arctic

EDITORS' NOTE   Conan Doyle's article appeared in *The Idler* for July 1892, the first year of a popular magazine started by his friends Robert Barr and Jerome K. Jerome, and later in America in *McClure's Magazine* (in which Conan Doyle had a financial stake) for March 1894. By this time well known as the creator of Sherlock Holmes, he lent his prestige to views about the Arctic, and ideas about how to reach the North Pole, that he had raised locally in his Portsmouth Literary & Scientific Society talk of December 1893 when he was a struggling physician without a literary reputation. Some details from it were reused later in his 1924 autobiography *Memories and Adventures*.

IT IS A strange thing to think that there is a body of men in Great Britain, the majority of whom have never since their boyhood seen the corn in the fields. It is the case with the whale fishers of Peterhead. They begin their hard life very early as boys or ordinary seamen, and from that time onwards they leave home at the end of February, before the first shoots are above the ground, and return in September, when only the stubble remains to show where the harvest has been. I have seen and spoken with many an old whaling man to whom a bearded ear of corn was a thing to be wondered over, and preserved.

The trade which these men follow is old and honourable. There was a time when the Greenland seas were harried by the ships of many nations, when the Basques and Biscayans were the great fishers of whales, and when Dutchmen, men of the Hansatowns, Spaniards, and Britons all joined in the great blubber hunt. Then one by one, as national energy or industrial capital decreased, the various countries tailed off, until, in the earlier part of this century, Hull, Poole, and Liverpool were three leading whaling ports. But again the trade shifted its centre. Scoresby was the last of the great English captains, and from his time the industry has gone more and more North, until the whaling of Greenland waters came to be monopolised by Peterhead, which shares the sealing, however, with Dundee, and with a fleet from Norway. But now, alas! the whaling appears to be upon its last legs, the Peterhead ships are seeking new outlets in the Antarctic seas, and a historical training-school of brave and hardy seamen will soon be a thing of the past.

It is not that the present generation is less persistent and skilful than its predecessors, nor is it that the Greenland whale is in danger of becoming extinct, but the true reason appears to be that Nature, while depriving this unwieldy mass of blubber of any weapons, has given it in compensation a highly intelligent brain. That the whale entirely understands the mechanism of its own capture is beyond dispute. To swim backwards and forwards beneath a floe in the hope of cutting the rope against the sharp edge of the ice is a common device of the creature after being struck. By degrees, however, it has realised the fact that there are limits to the powers of its adversaries, and that by keeping far in among the icefields it may shake off the most intrepid of pursuers. Gradually the creature has deserted the open sea, and bored deeper and deeper among the ice barriers, until now, at last, it really appears to have reached inaccessible feeding grounds, and it is seldom, indeed, that the watcher in the crow's nest sees the high plume of spray, and the broad, black tail in the air which sets his heart a-thumping.

But if a man have the good fortune to be present at a "fall," and, above all, if he be, as I have been, in the harpooning and in the lancing boat, he has a taste of sport which it would be ill to match. To play a salmon is a royal game, but when your fish weighs more than a suburban villa, and is worth a clear two thousand pounds, when, too, your line is a thumb's thickness of manilla rope with fifty strands, every strand tested for 36 lbs., it dwarfs all other experiences. And the lancing too, when the creature is spent, and your boat pulls in to give it the *coup de grace* with cold steel, that is also exciting! A hundred tons of despair are churning the waters up into a red foam, two great black fins are rising and falling like the sails of a windmill, casting the boat into a shadow as they droop over it, but still the harpooner clings to the head, where no harm can come, and, with the wooden butt of the twelve-foot lance against his stomach, he presses it home until the long struggle is finished, and the black back rolls over to expose the livid, whitish surface beneath. Yet amid all the excitement – and no one who has not held an oar in such a scene can tell how exciting it is – one's sympathies lie with the poor hunted creature. The whale has a small eye, little larger than that of a bullock, but I cannot easily forget the mute expostulation which I read in one, as it dimmed over in death within hand's touch of me. What could it guess, poor creature, of laws of supply and demand, or how could it imagine that when Nature placed an elastic filter inside its mouth, and when man discovered that the plates of which it was composed were the most pliable and yet durable things in creation, its death-warrant was signed.

Of course, it is only the one species, and the very rarest species of whale, which is the object of the fishery. The common rorqual or finner, largest of creatures upon this planet, whisks its eighty feet of worthless tallow round the whaler without fear of any missile more dangerous than a biscuit. This, with its good-for-nothing cousin, the hunch-back whale, abounds in the Arctic seas, and I have seen their sprays upon a clear day shooting up along the horizon like the smoke from a busy factory. A stranger sight still is, when looking over the bulwarks into the clear water, to see far down where the green is turning to black the huge, flickering figure of a whale gliding under the ship. And then the strange grunting, soughing noise which they make as they come up, with something of the contented pig in it, and something of the wind in the chimney!

Contented they may well be, for the finner has no enemies, save an occasional sword-fish, and Nature, which in a humorous mood has in the case of the right whale affixed the smallest of gullets to the largest of creatures, has dilated the swallow of its less valuable brother, so that it can have a merry time among the herrings.

The gallant seaman, who in all the books stands in the prow of a boat, waving a harpoon over his head, with the line snaking out into the air behind him, is only to be found now in Paternoster Row. The Greenland seas have not known him for more than a hundred years, since first the obvious proposition was advanced that one could shoot both harder and more accurately than one could throw. Yet one clings to the ideals of one's infancy, and I hope that another century may have elapsed before the brave fellow disappears from the frontispieces, in which he still throws his outrageous weapon an impossible distance. The swivel gun, like a huge horse-pistol, with its great oakum wad, and 28 drams of powder, is a more reliable, but a far less picturesque, object.

But to aim with such a gun is an art in itself, as will be seen when one considers that the rope is fastened to the neck of the harpoon, and that as the missile flies the downward drag of this rope must seriously deflect it. So difficult is it to make sure of one's aim, that it is the etiquette of the trade to pull the boat right on to the creature, the prow shooting up its soft, gently-sloping side, and the harpooner firing straight down into its broad back, into which not only the four-foot harpoon, but ten feet of the rope behind it, will disappear. Then, should the whale cast its tail in the air, after the time-honoured fashion of the pictures, that boat would be in evil case,

but, fortunately, when frightened or hurt it does no such thing, but curls its tail up underneath it, like a cowed dog, and sinks like a stone. Then the bows splash back into the water, the harpooner hugs his own soul, the crew light their pipes and keep their legs apart while the line runs merrily down the middle of the boat and over the bows. There are two miles of it there, and a second boat will lie alongside to splice on if the first should run short, the end being always kept loose for that purpose.

And now occurs the one serious danger of whaling. The line has usually been coiled when it was wet, and as it runs out it is very liable to come in loops which whizz down the boat between the men's legs. A man lassoed in one of these nooses is gone, and fifty fathoms deep, before the harpooner has time to say "Where's Jock?" Or if it be the boat itself which is caught, then down it goes like a cork on a trout-line, and the man who can swim with a whaler's high boots on is a swimmer indeed. Many a whale has had a Parthian revenge in this fashion. Some years ago a man was whisked over with a bight of rope round his thigh. "Christ, man, Alec's gone!" shrieked the boat-steerer, heaving up his axe to cut the line. But the harpooner caught his wrist. "Na, na, mun," he cried, "the oil money'll be a good thing for the widdie." And so it was arranged, while Alec shot on upon his terrible journey.

That oil money is the secret of the frantic industry of these seamen, who, when they do find themselves taking grease aboard, will work day and night, though night is but an expression up there, without a thought of fatigue. For the secure pay of officers and men is low indeed, and it is only by their share of the profits that they can hope to draw a good cheque when they return. Even the new-joined boy gets his shilling in the ton, and so draws an extra five pounds when a hundred tons of oil are brought back. It is practical socialism, and yet a less democratic community than a whaler's crew could not be imagined. The captain rules the mates, the mates the harpooners, the harpooners the boat-steerers, the boat-steerers the line-coilers, and so on in a graduated scale which descends to the ordinary seaman, who, in turn, bosses it over the boys. Every one of these has his share of oil money, and it may be imagined what a chill blast of unpopularity blows around the luckless harpooner who, by clumsiness or evil chance, has missed his whale. Public opinion has a terrorising effect even in those little floating communities of fifty souls. I have known a grizzled harpooner burst into tears when he saw by his slack line that he had missed his mark, and Aberdeenshire seamen are not a very soft race either.

Though twenty or thirty whales have been taken in a single year in Greenland seas, it is probable that the great slaughter of last century has diminished their number until there are not more than a few hundreds in existence. I mean, of course, of the right whale, for the others, as I have said, abound. It is difficult to compute the numbers of a species which comes and goes over great tracks of water and among huge icefields, but the fact that the same whale is often pursued by the same whaler upon successive trips shows how limited their number must be. There was one, I remember, which was conspicuous through having a huge wart, the size and shape of a beehive, upon one of the flukes of its tail. "I've been after that fellow three times," said the captain, as we dropped our boats. "He got away in '61. In '67 we had him fast, but the

harpoon drew. In '76 a fog saved him. It's odds that we have him now!" I fancied that the betting lay rather the other way myself, and so it proved, for that warty tail is still thrashing the Arctic seas for all that I know to the contrary.

I shall never forget my own first sight of a right whale. It had been seen by the look-out on the other side of a small icefield, but had sunk as we all rushed on deck. For ten minutes we awaited its re-appearance, and I had taken my eyes from the place, when a general gasp of astonishment made me glance up, and there was the whale *in the air*. Its tail was curved just as a trout's is in jumping, and every bit of its glistening lead-coloured body was clear of the water. It was little wonder that I should be astonished, for the captain, after thirty voyages, had never seen such a sight. On catching it we discovered that it was very thickly covered with a red, crab-like parasite, about the size of a shilling, and we conjectured that it was the irritation of these creatures which had driven it wild. If a man had short, nail-less flippers, and a prosperous family of fleas upon his back, he would appreciate the situation.

When a fish, as the whalers will for ever call it, is taken, the ship gets alongside, and the creature is fixed head and tail in a curious and ancient fashion, so that by slacking or tightening the ropes, each part of the vast body can be brought uppermost. A whole boat may be seen inside the giant mouth, the men hacking with axes to slice away the ten-foot screens of bone, while others with sharp spades upon the back are cutting off the deep greatcoat of fat in which kindly Nature has wrapped up this most overgrown of her children. In a few hours all is stowed away in the tanks, and a red islet, with white projecting bones, lies alongside, and sinks like a stone when the ropes are loosed. Some years ago a man, still lingering upon the back, had the misfortune to have his foot caught between the creature's ribs at the instant when the tackles were undone. Some æons hence those two skeletons, the one hanging by the foot from the other, may grace the museum of a subtropical Greenland, or astonish the students of the Spitzbergen Institute of Anatomy.

Apart from sport, there is a glamour about those circumpolar regions which must affect everyone who has penetrated to them. My heart goes out to that old, grey-headed whaling captain who, having been left for an instant when at death's door, staggered off in his night gear, and was found by his nurses far from his house and still, as he mumbled, "pushing to the norrard." So an Arctic fox, which a friend of mine endeavoured to tame escaped, and was caught many months afterwards in a gamekeeper's trap in Caithness. It also was pushing norrard, though who can say by what strange compass it took its bearings? It is a region of purity, of white ice and of blue water, with no human dwelling within a thousand miles to sully the freshness of the breeze which blows across the icefields. And then it is a region of romance also. You stand on the very brink of the unknown, and every duck that you shoot bears pebbles in its gizzard which come from a land which the maps know not.

These whaling captains profess to see no great difficulty in reaching the Pole. Some little margin must be allowed, no doubt, for expansive talk over a pipe and a glass, but still there is a striking unanimity in their ideas. Briefly they are these.

What bars the passage of the explorer as he ascends between Greenland and Spitzbergen is that huge floating ice-reef which scientific explorers have called "the palæocrystic sea," and the whalers, with more expressive Anglo-Saxon, "the barrier." The ship which has picked its way among the great ice-floes finds itself, somewhere about the 81st degree, confronted by a single mighty wall extending right across from side to side, with no chink or creek up which she can push her bows. It is old ice, gnarled and rugged, and of an exceeding thickness, impossible to pass, and nearly impossible to travel over, so cut and jagged is its surface. Over this it was that the gallant Parry struggled with his sledges in 1827, reaching a latitude (about 82°.30, if my remembrance is correct) which for a long time was the record. As far as he could see this old ice extended right away to the Pole.

Such is the obstacle. Now for the whaler's view of how it may be surmounted.

This ice, they say, solid as it looks, is really a floating body, and at the mercy of the water upon which it rests. There is in those seas a perpetual southerly drift, which weakens the cohesion of the huge mass, and when, in addition to this, the prevailing winds happen to be from the North, the barrier is all shredded out, and great bays and gulfs appear in its surface. A brisk northerly wind, long continued, might at any time clear a road, and has, according to their testimony, frequently cleared a road by which a ship might slip through to the Pole. Whalers fishing as far North as the 82nd degree have in an open season seen no ice, and, more important still, no reflection of ice in the sky to the north of them. But they are in the service of a company, they are there to catch whales, and there is no adequate inducement to make them risk themselves, their vessels, and their cargoes, in a dash for the North.

The matter might be put to the test without trouble or expense. Take a stout wooden gunboat, short and strong, with engines as antiquated as you like, if they be but a hundred horse power. Man her with a sprinkling of Scotch and Shetland seamen from the Royal Navy, and let the rest of the crew be lads who must have a training cruise in any case. For the first few voyages carry a couple of experienced ice masters, in addition to the usual Naval officers. Put a man like Markham in command. Then send this ship every June or July to inspect the barrier, with strict orders to keep out of the heavy ice unless there were a very clear water-way. For six years she might go in vain. On the seventh you might have an open season, hard, northerly winds, and a clear sea. In any case no expense or danger is incurred, and there could be no better training for young seamen. They will find the Greenland seas in summer much more healthy and pleasant than the Azores or Madeira, to which they are usually dispatched. The whole expedition should be done in less than a month.

Singular incidents occur in those northern waters, and there are few old whalers who have not their queer yarn, which is sometimes of personal and sometimes of general interest. There is one which always appeared to me to deserve more attention than has ever been given to it. Some years ago, Captain David Gray, of the *Eclipse*, the *doyen* of the trade, and the representative, with his brothers John and Alec, of a famous family of whalers, was cruising far to the North when he saw a large bird flapping over the ice. A boat was dropped, the bird shot, and brought aboard, but no man there could say what manner of fowl it was. Brought home, it was at once identified as being a half-grown albatross, and now stands in the Peterhead Museum with a neat little label to that effect between its webbed feet.

Now the albatross is an Antarctic bird, and it is quite unthinkable that this solitary specimen flapped its way from the other end of the earth. It was young, and possibly giddy, but quite incapable of a wild outburst of that sort. What is the alternative? It must have been a *Southern* straggler from some breed of albatrosses further North. But if there is a different fauna further North, then there must be a climactic change there. Perhaps Kane was not so far wrong after all in his surmise of an open Polar sea. It may be that that flattening at the Poles of the earth, which always seemed to my child-hood's imagination to have been caused by the finger and thumb of the Creator, when He held up this little planet before He set it spinning, has a greater influence on climate than we have yet ascribed to them. But if so, how simple would the task of our exploring ship become when a wind from the North had made a rift in the barrier.

There is little land to be seen during the seven months of a whaling cruise. The strange solitary island of Jan Mayen may possibly be sighted, with its great snow-capped ex-volcano jutting up among the clouds. In the palmy days of the whale fishing the Dutch had a boiling station there, and now great stones with iron rings let into them and rusted anchors lie littered about in this absolute wilderness as a token of their former presence. Spitzbergen, too, with its black crags and its white glaciers, a dreadful looking place, may possibly be seen. I saw it myself for the first and last time in a sudden rift in the drifting wrack of a furious gale, and for me it stands as the very emblem of stern grandeur. And then towards the end of the season the whalers come South to the 72nd degree, and try to bore in towards the coast of Greenland, in the South-eastern corner, and if you then, at the distance of eighty miles, catch the least glimpse of the loom of the cliffs, then, if you are anything of a dreamer, you will have plenty of food for dreams, for this is the very spot where one of the most interesting questions in the world is awaiting a solution.

Of course, it is a commonplace that when Iceland was one of the centres of civilisation in Europe, the Icelanders budded off a colony upon Greenland, which throve and flourished, and produced sagas of its own, and waged war upon the Skraelings or Esquimaux, and generally sang and fought and drank in the bad old, full-blooded fashion. So prosperous did they become that they built them a cathedral, and sent to Denmark for a bishop, there being no protection for local industries at that time. The bishop, however, was prevented from reaching his see by some sudden climatic change which brought the ice down between Iceland and Greenland, and from that day (it was in the 14th century) to this no one has penetrated that ice, nor has it ever been ascertained what became of that ancient city, or of its inhabitants. Have they preserved some singular civilisation of their own, and are they still singing and drinking and fighting, and waiting for the bishop from over the seas? Or have they been destroyed by the hated Skraelings, or have they, as is more likely, amalgamated with them, and produced a race of tow-headed, large-limbed Esquimaux? We must wait until some Nansen turns his steps in that direction before we can tell. At present it is one of those interesting historical questions, like the fate of those Vandals who were driven by Belisarius into the interior of Africa, which are far better unsolved. When we know everything about this earth, the romance and the poetry will all have been wiped away from it. There is nothing so artistic as a haze.

There is a good deal which I had meant to say about bears, and about seals, and about sea-unicorns, and sword-fish, and all the interesting things which combine to throw that glamour over the Arctic; but, as the genial critic is fond of remarking, it has all been said very much better already. There is one side of the Arctic regions, however, which has never had due attention paid to it, and that is the medical and curative side. Davos Platz has shown what cold can do in consumption, but in the life-giving air of the Arctic Circle no noxious germ can live. The only illness of any consequence which ever attacks a whaler is an explosive bullet. It is a safe prophecy that before many years are past, steam yachts will turn to the North every summer, with a cargo of the weak-chested, and people will understand that Nature's ice-house is a more healthy place than her vapour-bath.

Life on a Greenland Whaler

EDITORS' NOTE   This account of Conan Doyle's appeared in *The Strand Magazine* for January 1897, and in America, with the sub-title "A Record of Personal Adventures in the Arctic Seas," in *McClure's Magazine* that March. Conan Doyle apparently wrote it without constant recourse to his diary from aboard the *Hope*, for there are some differences in the details set forth, especially remembering the *Hope* returning home in September when his diary indicates it was actually August, 1880. The article served as the basis for chapter four ("Whaling in the Arctic Ocean") of his 1924 autobiography *Memories and Adventures*.

IT HAS BEEN my good fortune to have an experience of a life which is already extinct, for although whale-ships, both English and American, still go to Davis' Strait, the Greenland fishing – that is, the fishing in the waters between Greenland and Spitzbergen – has been attended with such ill-fortune during the last ten years that it has now been abandoned. The *Hope* and the *Eclipse*, both of Peterhead, were the last two vessels which clung to an industry which was once so flourishing that it could support a fleet of a hundred sail; and it was in the *Hope*, under the command of the well-known whaler, John Gray, that I paid a seven months' visit to the Arctic Seas in the year 1880. I went in the capacity of surgeon, but as I was only twenty years of age when I started, and as my knowledge of medicine was that of an average third year's student, I have often thought that it was as well that there was no very serious call upon my services.

It came about in this way. One raw afternoon in Edinburgh, whilst I was sitting reading hard for one of those examinations which blight the life of a medical student, there entered to me a fellow-student with whom I had some slight acquaintance. The monstrous question which he asked drove all thought of my studies out of my head.

"Would you care," said he, "to start next week for a whaling cruise? You'll be surgeon, two pound ten a month and three shillings a ton oil money."

"How do you know I'll get the berth?" was my natural question.

"Because I have it myself. I find at this last moment that I can't go, and I want to get a man to take my place."

"How about an Arctic kit?"

"You can have mine."

In an instant the thing was settled,

and within a few minutes the current of my life had been deflected into a new channel.

In little more than a week I was in Peterhead, and busily engaged, with the help of the steward, in packing away my scanty belongings in the locker beneath my bunk on the good ship *Hope*. And this, my first appearance aboard the ship, was marked by an absurd incident. In my student days boxing was a favourite amusement of mine, for I had found that when reading hard one can compress more exercise into a short time in this way than in any other. Among my belongings therefore were two pairs of battered and discoloured gloves. Now, it chanced that the steward was a bit of a fighting man, so when my unpacking was finished, he, of his own accord, picked up the gloves and proposed that we should then and there have a bout. I don't know whether Jack Lamb still lives – but if he does I am sure that he remembers the incident. I can see him now, blue-eyed, yellow-bearded, short but deep-chested, with the bandy legs of a very muscular man. Our contest was an unfair one, for he was several inches shorter in the reach than I, and knew nothing about sparring, although I have no doubt he was a formidable person in a street row. I kept propping him off as he rushed at me, and at last, finding that he was determined to bore his way in, I had to hit him out with some severity. An hour or so afterwards, as I sat reading in the saloon, there was a murmur in the mate's berth, which was next door, and suddenly I heard the steward say, in loud tones of conviction: "So help me, Colin, he's the best surrr-geon we've had! He's blackened my e'e!" It was the first (and very nearly the last) testimonial that I ever received to my professional abilities.

He was a good fellow, the steward,

and as I look back at that long voyage, during when for seven months we never set our feet upon land, his kindly, open face is one of those of which I like to think. He had a very beautiful and sympathetic tenor voice, and many an hour have I listened to it, with its accompaniment of rattling plates and jingling knives, as he cleaned up the dishes in his pantry. He knew a great store of pathetic and sentimental songs, and it is only when you have not seen a woman's face for six months that you realize what sentiment means. When Jack trilled out "Her Bright Smile Haunts Me Still," or "Wait for Me at Heaven's Gate, Sweet Belle Mahone," he filled us all with a vague, sweet discontent, which comes back to me now as I think of it. As to his boxing, he practised with me every day, and became a formidable opponent – especially when there was a sea on, when, with his more experienced sea-legs, he could come charging down with the heel of the ship. He was a baker by trade, and I dare say Greenland is as much a dream to him now as it is to me.

There was one curious thing about the manning of the *Hope*. The man who signed on as first mate was a little, decrepit, broken fellow, absolutely incapable of performing the duties. The cook's assistant, on the other hand, was a giant of a man, red-bearded, bronzed, with huge limbs, and a voice of thunder. But the moment that the ship cleared the harbour the little, decrepit mate disappeared into the cook's galley, and acted as scullery-boy for the voyage, while the mighty scullery-boy walked aft and became chief mate. The fact was, that the one had the certificate, but was past sailoring, while the other could neither read nor write, but was as fine a seaman as ever lived; so, by an agreement to which everybody concerned was party,

First Mate Colin McLean.

they swapped their berths when they were at sea.

Colin McLean, with his six foot of stature, his erect, stalwart figure, and his fierce, red beard, pouring out from between the flaps of his sealing-cap, was an officer by natural selection, which is a higher title than that of a Board of Trade certificate. His only fault was that he was a very hot-blooded man, and that a little would excite him to a frenzy. I have a vivid recollection of an evening which I spent in dragging him off from the steward, who had imprudently made some criticism upon his way of attacking a whale which had escaped. Both men had had some rum, which had made the one argumentative and the other violent, and as we were all three seated in a space of about seven by four, it took some hard work to prevent bloodshed. Every now and then, just as I thought all danger was past, the steward would begin again with his fatuous, "No offence, Colin, but all I says is that if you had been a bit quicker on the fush – " I don't know how often this sentence was begun, but never once was it ended; for at the word "fush" Colin always seized him by the throat, and I Colin round the waist, and we struggled until we were all panting and exhausted. Then when the steward had recovered a little breath he would start that miserable sentence once more, and the "fush" would be the signal for another encounter. I really believe that if I had not been there the mate would have killed him, for he was quite the angriest man that I have ever seen.

There were fifty men upon our whaler, of whom half were Scotchmen and half Shetlanders, whom we picked up at Lerwick as we passed. The Shetlanders were the steadier and more tractable, quiet, decent, and soft-spoken; while the Scotch seamen were more likely to give trouble, but also more virile and of stronger character. The officers and harpooners were all Scotch, but as ordinary seamen, and especially as boatmen, the Shetlanders were as good as could be wished.

There was only one man on board who belonged neither to Scotland nor to Shetland, and he was the mystery of the ship. He was a tall, swarthy, dark-eyed man, with blue-black hair and beard, singularly handsome features, and a curious reckless sling of his shoulders when he walked. It was rumoured that he came from the South of England, and that he had fled thence to avoid the law. He made friends with no one, and spoke very seldom, but he was one of the smartest seamen in the ship. I could believe from his appearance that his temper was Satanic, and that the crime for which he was hiding may have been a bloody one. Only once he gave us a glimpse of his hidden fires. The cook – a very burly, powerful man – the little mate was only assistant – had a private store of rum, and treated himself so liberally to it that for three successive days the dinner of the crew was ruined. On the third day our silent outlaw approached the cook with a brass saucepan in his hand. He said nothing, but he struck the man such a frightful blow that his head flew through the bottom, and the sides of the pan were left dangling round his neck. The half-drunken and half-stunned cook talked of fighting, but he was soon made to feel that the sympathy of the ship was against him, so he reeled back grumbling to his duties, while the avenger relapsed into his usual moody indifference. We heard no further complaints about the cooking.

There are eight boats on board a whaler, but it is usual to send out only seven, for it takes six men each to man them, so that when the seven are out no one is left on board except the so-called "idlers," who have not signed to do seamen's work at all. It happened, however, that on board the *Hope* the "idlers" were an exceptionally active and energetic lot, so we volunteered to man the eighth boat, and we made it, in our own estimation at least, one of the most efficient, both in sealing and in whaling. The steward, the second engineer, the donkey-engine man, and I pulled the oars, with a red-headed Highlander for harpooner, and the handsome outlaw to steer. Our tally of seals stood as high as any; and in whaling we were once the harpooning and once the lancing boat, so our record was an excellent one. So congenial was the work to me, that Captain Gray was good enough to offer to make me harpooner as well as surgeon if I would come with him upon a second voyage, with power to draw the double day. It is as well that I refused, for the life is such a fascinating one that I could imagine that a man would find it more and more difficult to give it up.

Most of the crew are never called upon to do so, for they spend their whole lives in the same trade. There were men on board the *Hope* who had never seen corn growing, for from their boyhood they had always started for the whaling in March and returned in September.

One of the charms of the work is the gambling element which is inherent in it. Every man shares in the profit – so much for the captain, so much for the mate, so much for the seaman. If the voyage is successful, everyone is rich until another spring comes round. If the ship comes home clean, it means a starvation winter for all hands. The men do not need to be told to be keen. The shout from the crow's-nest which tells of the presence of a whale, and the rattle of the falls as the boats are cleared away, blend into one sound. The watch below rush up from their bunks with their clothes over their arms, and spring into their boats in that Arctic air, waiting for a chance later for finishing their toilet. Woe betide the harpooner or the boat-steerer who by any clumsiness has missed a fish! He has taken a five-pound note out of the pocket of every meanest hand upon the ship. Black is his welcome when he returns to his fellows.

What surprised me most in the Arctic regions was the rapidity with which you reach them. I had never realized that they lie at our very doors. I think that we were only four days out from Shetland when we were among the drift ice. I awoke of a morning to hear the bump, bump of the floating pieces against the side of the ship, and I went on deck to see the whole sea covered with them to the horizon. They were none of them large, but they lay so thick that a man might travel far by springing from one to the other. Their dazzling whiteness made the sea seem bluer by contrast, and with a blue sky above, and that glorious Arctic air in one's nostrils, it was a morning to remember. Once on one of the swaying, rocking pieces we saw a huge seal, sleek, sleepy, and imperturbable, looking up with the utmost assurance at the ship, as if it knew that the close time had still three weeks to run. Further on we saw on the ice the long, human-like prints of a bear. All this with the snowdrops of Scotland still fresh in our glasses in the cabin.

I have spoken about the close time, and I may explain that, by an agreement between the Norwegian and British Governments, the subjects of both nations are forbidden to kill a seal before the 3rd of April. The reason for this is, that the breeding season is in March, and if the mothers should be killed before the young are able to take care of themselves, the race would soon become extinct. For breeding purposes, the seals all come together at a variable spot, which is evidently pre-arranged among them, and as this place may be anywhere within many hundreds of square miles of floating ice, it is no easy matter for the fisher to find it. The means by which he sets about it are simple but ingenious. As the ship makes its way through the loose ice-streams, a school of seals is observed travelling through the water. Their direction is carefully taken by compass and marked upon the chart. An hour afterwards perhaps another school is seen. This is also marked. When these bearings have been taken several times, the various lines upon the chart are prolonged until they intersect. At this point, or near it, it is likely that the main pack of the seals will be found.

When you do come upon it, it is a wonderful sight. I suppose it is the largest assembly of creatures upon the face of the world – and this upon the open ice-fields hundreds of miles from Greenland coast. Somewhere between 71 deg. and 75 deg. is the rendezvous, and the longitude is even vaguer; but the seals have no difficulty in finding the address. From the crow's-nest at the top of the main-mast, one can see no end of them. On the furthest visible ice one can still see that sprinkling of pepper grains. And the young lie everywhere also, snow-white slugs, with a little black nose and large, dark eyes. Their half-human cries fill the air; and when you are sitting in the cabin of a ship which is in the heart of the seal-pack, you would think you were next door to a monstrous nursery.

The *Hope* was one of the first to find the seal-pack that year, but before the day came when hunting was allowed, we had a succession of strong gales, followed by a severe roll, which tilted the floating ice and launched the young seals prematurely into the water. And so, when the law at last allowed us to begin work, Nature had left us with very little work to do. However, at dawn upon the third, the ship's company took to the ice, and began to gather in its murderous harvest. It is brutal work, though not more brutal than that which goes on to supply every dinner-table in the country. And yet those glaring crimson pools upon the dazzling white of the ice-fields, under the peaceful silence of a blue Arctic sky, did seem a horrible intrusion. But an inexorable demand creates an inexorable supply, and the seals, by their death, help to give a living to the long line of seamen, dockers, tanners, curers, triers, chandlers, leather merchants, and oil-sellers, who stand between this annual butchery on the one hand, and the exquisite, with his soft leather boots, or the savant, using a delicate oil for his philosophical instruments, upon the other.

I have cause to remember that first

day of sealing on account of the adventures which befell me. I have said that a strong swell had arisen, and as this was dashing the floating ice together the captain thought it dangerous for an inexperienced man to venture upon it. And so, just as I was clambering over the bulwarks with the rest, he ordered me back and told me to remain on board. My remonstrances were useless, and at last, in the blackest of tempers, I seated myself upon the top of the bulwarks, with my feet dangling over the outer side, and there I nursed my wrath, swinging up and down with the roll of the ship. It chanced, however, that I was really seated upon a thin sheet of ice which had formed upon the wood, and so when the swell threw her over to a particularly acute angle, I shot off and vanished into the sea between two ice-blocks. As I rose, I clawed on to one of these, and soon scrambled on board again. The accident brought about what I wished, however, for the captain remarked that as I was bound to fall into the ocean in any case, I might just as well be on the ice as on the ship. I justified his original caution by falling in twice again during the day, and I finished it ignominiously by having to take to my bed while all my clothes were drying in the engine-room. I was consoled for my misfortunes by finding that they amused the captain to such an extent that they drove the ill-success of our sealing out of his head, and I had to answer to the name of "the great northern diver" for a long time thereafter. I had a narrow escape once through stepping backwards over the edge of a piece of floating ice while I was engaged in skinning a seal. I had wandered away from the others, and no one saw my misfortune. The face of the ice was so even that I had no purchase by which to pull myself up, and my body was rapidly becoming numb in the freezing water. At last, however, I caught hold of the hind flipper of the dead seal, and there was a kind of nightmare tug-of-war, the question being whether I should pull the seal off or pull myself on. At last, however, I got my knee over the edge and rolled on to it. I remember that my clothes were as hard as a suit of armour by the time I reached the ship, and that I had to thaw my crackling garments before I could change them.

This April sealing is directed against the mothers and young. Then, in May, the sealer goes further north; and about latitude 77 deg. or 78 deg. he comes upon the old male seals, who are by no means such easy victims. They are wary creatures, and it takes good long-range shooting to bag them. Then, in June, the sealing is over, and the ship bears away further north still, until in the 79th or 80th degree she is in the best Greenland whaling latitudes. There she remains for three months or so, and if she is fortunate she may bring back 300 or 400 per cent. to her owners, and a nice little purse full for every man of her ship's company. Or if her profits be more modest, she has at least afforded such sport that every other sport is dwarfed by the comparison.

It is seldom that one meets anyone who understands the value of a Greenland whale. A well-boned and large one as she floats is worth to-day something between two and three thousand pounds. This huge price is due to the value of whalebone, which is a very rare commodity, and yet is absolutely essential for some trade purposes. The price tends to rise steadily, for the number of the creatures is diminishing. In 1880, Captain Gray calculated that there were probably not more than 300 of them left alive in the whole expanse of the Greenland seas, an area of thousands of square miles. How few there are is shown by the fact that he recognized individuals amongst those which we chased. There was one with a curious wart about the size of a beehive upon his tail, which he had remembered chasing when he was a lad on his father's ship. Perhaps other generations of whalers may follow that warty tail, for the whale is a very long-lived creature. How long they live has never been ascertained; but in the days when it was customary to stamp harpoons with the names of vessels, old harpoons have been cut out of whales bearing names long forgotten in the trade, and all the evidence goes to prove that a century is well within their powers.

It is exciting work pulling on to a whale. Your own back is turned to him, and all you know about him is what you read upon the face of the boat-steerer. He is staring out over your head, watching the creature as it swims slowly through the water, raising his hand now and again as a signal to stop rowing when he sees that the eye is coming round, and then resuming the stealthy approach when the whale is end on. There are so many floating pieces of ice, that as long as the oars are quiet the boat alone will not cause the creature to dive. So you creep slowly up, and at last you are so near that the boat-steerer knows that you can get there before the creature has time to dive – for it takes some little time to get that huge body into motion. You see a sudden gleam in his eyes, and a flush in his cheeks, and it's "Give way, boys! Give way, all! Hard!" Click goes the trigger of the big harpoon gun, and the foam flies from your oars. Six strokes, perhaps, and then with a dull, greasy squelch the bows run upon something soft, and you and your oars are sent flying in every direction. But little you

care for that, for as you touched the whale you have heard the crash of the gun, and know that the harpoon has been fired point-blank into the huge, lead-coloured curve of its side. The creature sinks like a stone, the bows of the boat splash down into the water again, but there is the little red Jack flying from the centre thwart to show that you are fast, and there is the line whizzing swiftly under the seats and over the bows between your outstretched feet.

And there is the one element of danger – for it is rarely indeed that the whale has spirit enough to turn upon its enemies. The line is very carefully coiled by a special man named the line-coiler, and it is warranted not to kink. If it should happen to do so, however, and if the loop catches the limbs of any one of the boat's crew, that man goes to his death so rapidly that his comrades hardly know that he has gone. It is a waste of fish to cut the line, for the victim is already hundreds of fathoms deep.

"Haud your hand, mon," cried the harpooner, as a seaman raised his knife on such an occasion. "The fush will be a fine thing for the widdey." It sounds callous, but there was philosophy at the base of it.

This is the harpooning, and that boat has no more to do. But the lancing, when the weary fish is killed with the cold steel, is a more exciting because it is a more prolonged experience. You may be for half an hour so near to the creature that you can lay your hand upon its slimy side. The whale appears to have but little sensibility to pain, for it never winces when the long lances are passed through its body. But its instinct urges it to get its tail to work on the boats, and yours urges you to keep poling and boat-hooking along its side, so as to retain your safe position near its shoulder. Even there, however, we found upon this occasion that we were not quite out of danger's way, for the creature in its flurry raised its huge side-flapper and poised it over the boat. One flap would have sent us to the bottom of the sea, and I can never forget how, as we pushed our way from under, each of us held one hand up to stave off that great, threatening fin – as if any strength of ours could have availed if the whale had meant it to descend. But it was spent with loss of blood, and instead of coming down the fin rolled over the other way, and we knew that it was dead. Who would swap that moment for any other triumph that sport can give?

The peculiar other-world feeling of the Arctic regions – a feeling so singular, that if you have once been there the thought of it haunts you all your life – is due largely to the perpetual daylight. Night seems more orange-tinted and subdued than day, but there is no great difference. Some captains have been known to turn their hours right round out of caprice, with breakfast at night and supper at ten in the morning. There are your twenty-four hours, and you may carve them as you like. After a month or two the eyes grow weary of the eternal light, and you appreciate what a soothing thing our darkness is. I can remember

as we came abreast of Iceland, on our return, catching our first glimpse of a star, and being unable to take my eyes from it, it seemed such a dainty little twinkling thing. Half the beauties of Nature are lost through over-familiarity.

Your sense of loneliness also heightens the effect of the Arctic Seas. When we were in whaling latitudes it is probable that, with the exception of our consort, there was no vessel within 800 miles of us. For seven long months no letter and no news came to us from the southern world. We had left in exciting times. The Afghan campaign had been undertaken, and war seemed imminent with Russia. We returned opposite the mouth of the Baltic without any means of knowing whether some cruiser might not treat us as we had treated the whales. When we met a fishing-boat at the north of Shetland our first inquiry was as to peace or war. Great events had happened during those seven months: the defeat of Maiwand and the famous march of Roberts from Cabul to Candahar. But it was all haze to us; and, to this day, I have never been able to get that particular bit of military history straightened out in my own mind.

The perpetual light, the glare of the white ice, the deep blue of the water, these are the things which one remembers most clearly, and the dry, crisp, exhilarating air, which makes mere life the keenest of pleasures. And then there are the innumerable sea-birds, whose call is for ever ringing in your ears – the gulls, the fulmars, the snow-birds, the burgomasters, the loons, and the rotjes. These fill the air, and below, the waters are for ever giving you a peep of some strange new creature. The commercial whale may not often come your way, but his less valuable brethren abound on every side. The finner shows his ninety feet of worthless tallow, with the absolute conviction that no whaler would condescend to lower a boat for him. The mis-shapen hunchback whale, the ghost-like white whale, the narwhal, with his unicorn horn, the queer-looking bottle-nose, the huge, sluggish, Greenland shark, and the terrible killing grampus, the most formidable of all the monsters of the deep, these are the creatures who own those unsailed seas. On the ice are the seals, the saddle-backs, the ground seals and the huge bladdernoses, 12 ft. from nose to tail, with the power of blowing up a great blood-red football upon their noses when they are angry, which they usually are. Occasionally one sees a white Arctic fox upon the ice, and everywhere are the bears. The floes in the neighbourhood of the sealing-ground are all criss-crossed with their tracks – poor harmless creatures, with the lurch and roll of a deep-sea mariner. It is for the sake of the seals that they come out over those hundreds of miles of ice – and they have a very ingenious method of catching them, for they will choose a big ice-field with just one blow-hole for seals in the middle of it. Here the bear will squat, with its powerful forearms crooked round the hole. Then, when the seal's head pops up, the great paws snap together, and Bruin has got his luncheon. We used occasionally to burn some of the cook's refuse in the engine-room fires, and the smell would, in a few hours, bring up every bear for many miles to leeward of us.

But pleasant as the voyage is, there comes a day when the prow must be turned south once more. The winter comes on very suddenly sometimes, and woe betide the whaler which may be caught lagging. In September, then, our boats were taken in, our blubber tanks screwed down, and the *Hope* was fairly homeward bound. Far off loomed the huge peak of Jan-Mayen Island, the ice-blink glimmered and faded away behind us, and I had seen the last which I am ever, save in my dreams, likely to see of the Greenland Ocean.

The Captain of the "Pole-Star"

EDITORS' NOTE First published in *Temple Bar* magazine in January of 1883, "The Captain of the Pole-Star" marked an important early success for Conan Doyle. Drawing heavily on his experiences aboard the *Hope*, and taking the form of a journal kept by a medical student serving aboard a whaling ship as its surgeon, it demonstrated how this "strange and fascinating chapter" of his life had fired his imagination. *Temple Bar* paid him an impressive ten guineas for the tale; Conan Doyle wrote to a friend, "I think they liked it," and his name began to become known to British readers as a promising new short-story writer.

[Being an extract from the singular journal of John M'Alister Ray, student of medicine.]

September 11th. – Lat. 81° 40′ N.; long. 2° E. Still lying-to amid enormous ice-fields. The one which stretches away to the north of us, and to which our ice-anchor is attached, cannot be smaller than an English county. To the right and left unbroken sheets extend to the horizon. This morning the mate reported that there were signs of pack ice to the southward. Should this form of sufficient thickness to bar our return, we shall be in a position of danger, as the food, I hear, is already running somewhat short. It is late in the season, and the nights are beginning to reappear. This morning I saw a star twinkling just over the fore-yard, the first since the beginning of May. There is considerable discontent among the crew, many of whom are anxious to get back home to be in time for the herring season, when labour always commands a high price upon the Scotch coast. As yet their displeasure is only signified by sullen countenances and black looks, but I heard from the second mate this afternoon that they contemplated sending a deputation to the captain to explain their grievance. I much doubt how he will receive it, as he is a man of fierce temper, and very sensitive about anything approaching to an infringement of his rights. I shall venture after dinner to say a few words to him upon the subject. I have always found that he will tolerate from me what he would resent from any other member of the crew. Amsterdam Island, at the north-west corner of Spitzbergen, is visible upon our starboard quarter – a rugged line of volcanic rocks, intersected by white seams, which represent glaciers. It is curious to think that at the present moment there is probably no human being nearer to us than the Danish settlements in the south of Greenland – a good nine hundred miles as the crow flies. A captain takes a great responsibility upon himself when he risks his vessel under such circumstances. No whaler has ever remained in these latitudes till so advanced a period of the year.

9 P.M. – I have spoken to Captain Craigie, and though the result has been hardly satisfactory, I am bound to say that he listened to what I had to say very quietly and even deferentially. When I had finished he put on that air of iron determination which I have frequently observed upon his face, and paced rapidly backwards and forwards across the narrow cabin for some minutes. At first I feared that I had seriously offended him, but he dispelled the idea by sitting down again, and putting his hand upon my arm with a gesture which almost amounted to a caress. There was a depth of tenderness too in his wild dark eyes which surprised me considerably.

"Look here, Doctor," he said, "I'm sorry I ever took you – I am indeed – and I would give fifty pounds this minute to see you standing safe upon the Dundee quay. It's hit or miss with me this time. There are fish to the north of us. How dare you shake your head, sir, when I tell you I saw them blowing from the mast-head?" – this in a sudden burst of fury, though I was not conscious of having shown any signs of doubt. "Two-and-twenty fish in as many minutes as I am a living man, and not one under ten foot.* Now, Doctor, do you think I can leave the country when there is only one infernal strip of ice between me and my fortune? If it came on to blow from the

---

* A whale is measured among whalers not by the length of its body, but by the length of its whalebone.

north to-morrow we could fill the ship and be away before the frost could catch us. If it came on to blow from the south – well, I suppose the men are paid for risking their lives, and as for myself it matters but little to me, for I have more to bind me to the other world than to this one. I confess that I am sorry for you, though. I wish I had old Angus Tait who was with me last voyage, for he was a man that would never be missed, and you – you said once that you were engaged, did you not?"

"Yes," I answered, snapping the spring of the locket which hung from my watch-chain, and holding up the little vignette of Flora.

"Curse you!" he yelled, springing out of his seat, with his very beard bristling with passion. "What is your happiness to me? What have I to do with her that you must dangle her photograph before my eyes?" I almost thought that he was about to strike me in the frenzy of his rage, but with another imprecation he dashed open the door of the cabin and rushed out upon deck, leaving me considerably astonished at his extraordinary violence. It is the first time that he has ever shown me anything but courtesy and kindness. I can hear him pacing excitedly up and down overhead as I write these lines.

I should like to give a sketch of the character of this man, but it seems presumptuous to attempt such a thing upon paper, when the idea in my own mind is at best a vague and uncertain one. Several times I have thought that I grasped the clue which might explain it, but only to be disappointed by his presenting himself in some new light which would upset all my conclusions. It may be that no human eye but my own shall ever rest upon these lines, yet as a psychological study I shall attempt to leave some record of Captain Nicholas Craigie.

A man's outer case generally gives some indication of the soul within. The captain is tall and well-formed, with dark, handsome face, and a curious way of twitching his limbs, which may arise from nervousness, or be simply an outcome of his excessive energy. His jaw and whole cast of countenance is manly and resolute, but the eyes are the distinctive feature of his face. They are of the very darkest hazel, bright and eager, with a singular mixture of recklessness in their expression, and of something else which I have sometimes thought was more allied with horror than any other emotion. Generally the former predominated, but on occasions, and more particularly when he was thoughtfully inclined, the look of fear would spread and deepen until it imparted a new character to his whole countenance. It is at these times that he is most subject to tempestuous fits of anger, and he seems to be aware of it, for I have known him lock himself up so that no one might approach him until his dark hour was passed. He sleeps badly, and I have heard him shouting during the night, but his cabin is some little distance from mine, and I could never distinguish the words which he said.

This is one phase of his character, and the most disagreeable one. It is only through my close association with him, thrown together as we are day after day, that I have observed it. Otherwise he is an agreeable companion, well-read and entertaining, and as gallant a seaman as ever trod a deck. I shall not easily forget the way in which he handled the ship when we were caught by a gale among the loose ice at the beginning of April. I have never seen him so cheerful, and even hilarious, as he was that night, as he paced backwards and forwards upon the bridge amid the flashing of the lightning and the howling of the wind. He has told me several times that the thought of death was a pleasant one to him, which is a sad thing for a young man to say; he cannot be much more than thirty, though his hair and moustache are already slightly grizzled. Some great sorrow must have overtaken him and blighted his whole life. Perhaps I should be the same if I lost my Flora – God knows! I think if it were not for her that I should care very little whether the wind blew from the north or the south to-morrow. There, I hear him come down the companion, and he has locked himself up in his room, which shows that he is still in an unamiable mood. And so to bed, as old Pepys would say, for the candle is burning down (we have to use them now since the nights are closing in), and the steward has turned in, so there are no hopes of another one.

September 12th. – Calm, clear day, and still lying in the same position. What wind there is comes from the south-east but it is very slight. Captain is in a better humour, and apologised to me at breakfast for his rudeness. He still looks somewhat distrait, however, and retains that wild look in his eyes which in a Highlander would mean that he was "fey" – at least so our chief engineer remarked to me, and he has some reputation among the Celtic portion of our crew as a seer and expounder of omens.

It is strange that superstition should have obtained such mastery over this hard-headed and practical race. I could not have believed to what an extent it is carried had I not observed it for myself. We have had a perfect epidemic of it this voyage, until I have felt inclined to serve out rations of sedatives and nerve-tonics with the Saturday allowance of grog. The first symptom of it was that shortly after leaving Shetland the men at the wheel used to complain that they heard plaintive cries and screams in the wake of the ship, as if something were following it and were unable to overtake it. This fiction has been kept up during the whole voyage, and on dark nights at the beginning of the seal-fishing it was only with great difficulty that men could be induced to do their spell. No doubt what they heard was either the creaking of the rudder-chains, or the cry of some passing sea-bird. I have been fetched out of bed several times to listen to it, but I need hardly say that I was never able to distinguish anything unnatural. The men, however, are so absurdly positive upon the subject that it is hopeless to argue with them. I mentioned the matter to the captain once, but to my surprise he took it very gravely, and indeed appeared to be considerably disturbed by what I told him. I should have thought that he at least would have been above such vulgar delusions.

All this disquisition upon superstition leads me up to the fact that Mr. Manson, our second mate, saw a ghost last night – or, at least, says that he did, which of course is the same thing. It is quite refreshing to have some new topic of conversation after the eternal routine of bears and whales which has served us for so many months. Manson swears the ship is haunted; and that he would not stay in her a day if he had any other place to go to. Indeed the fellow is honestly frightened, and I had to give him some chloral and bromide of potassium this morning to steady him down. He seemed quite indignant when I suggested that he had been having an extra glass the night before, and I was obliged to pacify him by keeping as grave a countenance as possible during his story, which he

certainly narrated in a very straightfor- ward and matter-of-fact way.

"I was on the bridge," he said, "about four bells in the middle watch, just when the night was at its darkest. There was a bit of a moon, but the clouds were blowing across it so that you couldn't see far from the ship. John M'Leod, the harpooner, came aft from the fo'c'sle- head and reported a strange noise on the starboard bow. I went forrard and we both heard it, sometimes like a bairn crying and sometimes like a wench in pain. I've been seventeen years to the country and I never heard seal, old or young, make a sound like that. As we were standing there on the fo'c'sle-head the moon came out from behind a cloud, and we both saw a sort of white figure moving across the ice-field in the same direction that we had heard the cries. We lost sight of it for a while, but it came back on the port bow, and we could just make it out like a shadow on the ice. I sent a hand aft for the rifles, and M'Leod and I went down on to the pack, thinking that maybe it might be a bear. When we got on the ice I lost sight of M'Leod, but I pushed on in the direction where I could still hear the cries. I followed them for a mile or maybe more, and then running round a hummock I came right on to the top of it standing and waiting for me seemingly. I don't know what it was. It wasn't a bear, anyway. It was tall and white and straight, and if it wasn't a man nor a woman, I'll stake my davy it was something worse. I made for the ship as hard as I could run, and precious glad I was to find myself aboard. I signed articles to do my duty by the ship, and on the ship I'll stay, but you don't catch me on the ice again after sundown."

That is his story, given as far as I can in his own words. I fancy what he saw must, in spite of his denial, have been a young bear erect upon its hind legs, an attitude which they often assume when alarmed. In the uncertain light this would bear a resemblance to a human figure, especially to a man whose nerves were already somewhat shaken. Whatever it may have been, the occurrence is unfortunate, for it has produced a most unpleasant effect upon the crew. Their looks are more sullen than before, and their discontent more open. The double grievance of being debarred from the herring fishing and of being detained in what they choose to call a haunted vessel, may lead them to do something rash. Even the harpooners, who are the oldest and steadiest among them, are joining in the general agitation.

Apart from this absurd outbreak of superstition, things are looking rather more cheerful. The pack which was forming to the south of us has partly cleared away, and the water is so warm as to lead me to believe that we are lying in one of those branches of the gulf-stream which run up between Greenland and Spitzbergen. There are numerous small Medusæ and sea-lemons about the ship, with abundance of shrimps, so that there is every possibility of "fish" being sighted. Indeed one was seen blowing about dinner-time, but in such a position that it was impossible for the boats to follow it.

September 13th. – Had an interesting conversation with the chief mate, Mr. Milne, upon the bridge. It seems that our captain is as great an enigma to the seamen, and even to the owners of the vessel, as he has been to me. Mr. Milne tells me that when the ship is paid off, upon returning from a voyage, Captain Craigie disappears, and is not seen again until the approach of another season, when he walks quietly into the office of the company, and asks whether his services will be required. He has no friend in Dundee, nor does anyone pretend to be acquainted with his early history. His position depends entirely upon his skill as a seaman, and the name for courage and coolness which he had earned in the capacity of mate, before being entrusted with a separate command. The unanimous opinion seems to be that he is not a Scotchman, and that his name is an assumed one. Mr. Milne thinks that he has devoted himself to whaling simply for the reason that it is the most dangerous occupation which he could select, and that he courts death in every possible manner. He mentioned several instances of this, one of which is rather curious, if true. It seems that on one occasion he did not put in an appearance at the office, and a substitute had to be selected in his place. That was at the time of the last Russian and Turkish War. When he turned up again next spring he had a puckered wound in the side of his neck which he used to endeavour to conceal with his cravat. Whether the mate's inference that he had been engaged in the war is true or not I cannot say. It was certainly a strange coincidence.

The wind is veering round in an easterly direction, but is still very slight. I think the ice is lying closer than it did yesterday. As far as the eye can reach on every side there is one wide expanse of spotless white, only broken by an occasional rift or the dark shadow of a hummock. To the south there is the narrow lane of blue water which is our sole means of escape, and which is closing up every day. The captain is taking a heavy responsibility upon himself. I hear that the tank of potatoes has been finished, and even the biscuits are running short, but he preserves the same impassable countenance, and spends the greater part of the day at the crow's nest, sweeping the horizon with his glass. His manner is very variable, and he seems to avoid my society, but there has been no repetition of the violence which he showed the other night.

7.30 P.M. – My deliberate opinion is that we are commanded by a madman. Nothing else can account for the extra-ordinary vagaries of Captain Craigie. It is fortunate that I have kept this journal of our voyage, as it will serve to justify us in case we have to put him under any sort of restraint, a step which I should only consent to as a last resource. Curiously enough it was he himself who suggested lunacy and not mere eccentric-ity as the secret of his strange conduct. He was standing upon the bridge about an hour ago, peering as usual through his glass, while I was walking up and down the quarter-deck.

The majority of the men were below at their tea, for the watches have not been regularly kept of late. Tired of walking, I leaned against the bulwarks, and admired the mellow glow cast by the sinking sun upon the great ice-fields which surround us. I was suddenly aroused from the reverie into which I had fallen by a hoarse voice at my elbow, and starting round I found that the captain had descended and was standing by my side. He was staring out over the ice with an expression in which horror, surprise, and something approaching to joy were contending for the mastery. In spite of the cold, great drops of perspiration were coursing down his forehead, and he was evidently fearfully excited. His limbs twitched like those of a man upon the verge of an epileptic fit, and the lines about his mouth were drawn and hard.

"Look!" he gasped, seizing me by the wrist, but still keeping his eyes upon the distant ice, and moving his head slowly in a horizontal direction, as if following some object which was moving across the field of vision. "Look! There, man, there! Between the hummocks! Now coming out from behind the far one! You see her – you *must* see her! There still! Flying from me, by God, flying from me – and gone!"

He uttered the last two words in a whisper of concentrated agony which shall never fade from my remembrance. Clinging to the ratlines he endeavoured to climb up upon the top of the bulwarks as if in the hope of obtaining a last glance at the departing object. His strength was not equal to the attempt, however, and he staggered back against the saloon skylights, where he leaned panting and exhausted. His face was so livid that I expected him to become unconscious, so lost no time in leading him down the companion, and stretching him upon one of the sofas in the cabin. I then poured him out some brandy, which I held to his lips, and which had a wonderful effect upon him, bringing the blood back into his white face and steadying his poor shaking limbs. He raised himself up upon his elbow, and looking round to see that we were alone he beckoned to me to come and sit beside him.

"You saw it, didn't you?" he asked, still in the same subdued awesome tone so foreign to the nature of the man.

"No, I saw nothing."

His head sank back again upon the cushions. "No, he wouldn't without the glass," he murmured. "He couldn't. It was the glass that showed her to me, and then the eyes of love – the eyes of love. I say, Doc, don't let the steward in! He'll think I'm mad. Just bolt the door, will you!"

I rose and did what he had commanded.

He lay quiet for a while, lost in thought apparently, and then raised himself up upon his elbow again, and asked for some more brandy.

"You don't think I am, do you, Doc?" he asked, as I was putting the bottle back into the after-locker. "Tell me now, as man to man, do you think that I am mad?"

"I think you have something on your mind," I answered, "which is exciting you and doing you a good deal of harm."

"Right there, lad!" he cried, his eyes sparkling from the effects of the brandy. "Plenty on my mind – plenty! But I can work out the latitude and the longitude, and I can handle my sextant and manage my logarithms. You couldn't prove me mad in a court of law, could you, now?" It was curious to hear the man lying back and coolly arguing out the question of his own sanity.

"Perhaps not," I said; "but still I think you would be wise to get home as soon as you can, and settle down to a quiet life for a while."

"Get home, eh?" he muttered, with a sneer upon his face. "One word for me and two for yourself, lad. Settle down with Flora – pretty little Flora. Are bad dreams signs of madness?"

"Sometimes," I answered.

"What else? What would be the first symptoms?"

"Pains in the head, noises in the ears, flashes before the eyes, delusions –"

"Ah! what about them?" he interrupted. "What would you call a delusion?"

"Seeing a thing which is not there is a delusion."

"But she *was* there!" he groaned to himself. "She *was* there!" and rising, he unbolted the door and walked with slow and uncertain steps to his own cabin, where I have no doubt that he will remain until to-morrow morning. His system seems to have received a terrible shock, whatever it may have been that he imagined himself to have seen. The man becomes a greater mystery every day, though I fear that the solution which he has himself suggested is the correct one, and that his reason is affected. I do not think that a guilty conscience has anything to do with his behaviour. The idea is a popular one among the officers, and, I believe, the crew; but I have seen nothing to support it. He has not the air of a guilty man, but of one who has had terrible usage at the hands of fortune, and who should be regarded as a martyr rather than a criminal.

The wind is veering round to the south to-night. God help us if it blocks that narrow pass which is our only road to safety! Situated as we are on the edge of the main Arctic pack, or the "barrier" as it is called by the whalers, any wind from the north has the effect of shredding out the ice around us and allowing our escape, while a wind from the south blows up all the loose ice behind us, and hems us in between two packs. God help us, I say again!

September 14th. – Sunday, and a day of rest. My fears have been confirmed, and the thin strip of blue water has disappeared from the southward. Nothing but the great motionless ice-fields around us, with their weird hummocks and fantastic pinnacles. There is a deathly silence over their wide expanse which is horrible. No lapping of the waves now, no cries of seagulls or straining of sails, but one deep universal silence in which the murmurs of the seamen, and the creak of their boots upon the white shining deck, seem discordant and out of place. Our only visitor was an Arctic fox, a rare animal upon the pack, though common enough upon the land. He did not come near the ship, however, but after surveying us from a distance fled rapidly across the ice. This was curious conduct, as they generally know nothing of man, and being of an inquisitive nature, become so familiar that they are easily captured. Incredible as it may seem, even this little incident produced a bad effect upon the crew. "Yon puir beastie kens mair, ay, an' sees mair nor you nor me!" was the comment of one of the leading harpooners, and the others nodded their acquiescence. It is vain to attempt to argue against such puerile superstition. They have made up their minds that there is a curse upon the ship, and nothing will ever persuade them to the contrary.

The captain remained in seclusion all day except for about half an hour in the afternoon, when he came out upon the quarter-deck. I observed that he kept his eye fixed upon the spot where the vision of yesterday had appeared, and was quite prepared for another outburst, but none such came. He did not seem to see me, although I was standing close beside him. Divine service was read as usual by the chief engineer. It is a curious thing that in whaling vessels the Church of England Prayer-book is always employed, although there is never a member of that Church among either officers or crew. Our men are all Roman Catholics or Presbyterians, the former predominating. Since a ritual is used which is foreign to both, neither can complain that the other is preferred to them, and they listen with all attention and devotion, so that the system has something to recommend it.

A glorious sunset, which made the great fields of ice look like a lake of blood. I have never seen a finer and at the same time more weird effect. Wind is veering round. If it will blow twenty-four hours from the north all will yet be well.

September 15th. – To-day is Flora's birthday. Dear lass! it is well that she cannot see her boy, as she used to call me, shut up among the ice-fields with a crazy captain and a few weeks' provisions. No doubt she scans the shipping list in the *Scotsman* every morning to see if we are reported from Shetland. I have to set an example to the men and look cheery and unconcerned; but God knows, my heart is very heavy at times.

The thermometer is at nineteen Fahrenheit to-day. There is but little wind, and what there is comes from an unfavourable quarter. Captain is in an excellent humour; I think he imagines he has seen some other omen or vision, poor fellow, during the night, for he came into my room early in the morning, and stooping down over my bunk, whispered, "It wasn't a delusion, Doc; it's all right!" After breakfast he asked me to find out how much food was left, which the second mate and I proceeded to do. It is even less than we had expected. Forward they have half a tank full of biscuits, three barrels of salt meat, and a very limited supply of coffee beans and sugar. In the after-hold and lockers there are a good many luxuries, such as tinned salmon, soups, haricot mutton, etc., but they will go a very short way among a crew of fifty men. There are two barrels of flour in the store-room, and an unlimited supply of tobacco. Altogether there is about enough to keep the men on half rations for eighteen or twenty days – certainly not more. When we reported the state of things to the captain, he ordered all hands to be piped, and addressed them from the quarter-deck. I never saw him to better advantage. With his tall, well-knit figure, and dark animated face, he seemed a man born to command, and he discussed the situation in a cool sailor-like way which showed that while appreciating the danger he had an eye for every loophole of escape.

"My lads," he said, "no doubt you think I brought you into this fix, if it is a fix, and maybe some of you feel bitter against me on account of it. But you must remember that for many a season no ship that comes to the country has brought in as much oil-money as the old *Pole-Star*, and every one of you has had his share of it. You can leave your wives behind you in comfort, while other poor fellows come back to find their lassies on the parish. If you have to thank me for the one you have to thank me for the other, and we may call it quits. We've tried a bold venture before this and succeeded, so now that we've tried one and failed we've no cause to cry out about it. If the worst comes to the worse, we can make the land across the ice, and lay in a stock of seals which will keep us alive until the spring. It won't come to that, though, for you'll see the Scotch coast again before three weeks are out. At present every man must go on half rations, share and share alike, and no favour to any. Keep up your hearts and you'll pull through this as you've pulled through many a danger before." These few simple words of his had a wonderful effect upon the crew. His former unpopularity was forgotten, and the old harpooner whom I have already mentioned for his superstition, led off three cheers, which were heartily joined in by all hands.

September 16th. – The wind has veered round to the north during the night, and the ice shows some symptoms of opening out. The men are in a good humour in spite of the short allowance upon which they have been placed. Steam is kept up in the engine-room, that there may be no delay should an opportunity for escape present itself. The captain is in exuberant spirits, though he still retains that wild "fey" expression which I have already remarked upon. This burst of cheerfulness puzzles me more than his former gloom. I cannot understand it. I think I mentioned in an early part of this journal that one of his oddities is that he never permits any person to enter his cabin, but insists upon making his own bed, such as it is, and performing every other office for himself. To my surprise he handed me the key to-day and requested me to go down there and take the time by his chronometer while he measured the altitude of the sun at noon. It is a bare little room, containing a washing-stand and a few books, but little else in the way of luxury, except some pictures upon the walls. The majority of these are small cheap oleographs, but there was one water-colour sketch of the head of a young lady which arrested my attention. It was evidently a portrait, and not one of those fancy types of female beauty which sailors particularly affect. No artist could have evolved from his own mind such a curious mixture of character and weakness. The languid, dreamy eyes, with their drooping lashes, and the broad, low brow, unruffled by thought or care, were in strong contrast with the clean-cut, prominent jaw, and the resolute set of the lower lip. Underneath it in one of the corners was written, "M.B., æt. 19." That anyone in the short space of nineteen years of existence could develop such strength of will as was stamped upon her face seemed to me at the time to be well-nigh incredible. She must have been an extraordinary woman. Her features have thrown such a glamour over me that, though I had but a fleeting glance at them, I could, were

I a draughtsman, reproduce them line for line upon this page of the journal. I wonder what part she has played in our captain's life. He has hung her picture at the end of his berth, so that his eyes continually rest upon it. Were he a less reserved man I should make some remark upon the subject. Of the other things in his cabin there was nothing worthy of mention – uniform coats, a camp-stool, small looking-glass, tobacco-box, and numerous pipes, including an oriental hookah – which, by the by, gives some colour to Mr. Milne's story about his participation in the war, though the connection may seem rather a distant one.

11.20 P.M. – Captain just gone to bed after a long and interesting conversation on general topics. When he chooses he can be a most fascinating companion, being remarkably well-read, and having the power of expressing his opinion forcibly without appearing to be dogmatic. I hate to have my intellectual toes trod upon. He spoke about the nature of the soul, and sketched out the views of Aristotle and Plato upon the subject in a masterly manner. He seems to have a leaning for metempsychosis and the doctrines of Pythagoras. In discussing them we touched upon modern spiritualism, and I made some joking allusion to the impostures of Slade, upon which, to my surprise, he warned me most impressively against confusing the innocent with the guilty, and argued that it would be as logical to brand Christianity as an error because Judas, who professed that religion, was a villain. He shortly afterwards bade me good night and retired to his room.

The wind is freshening up, and blows steadily from the north. The nights are as dark now as they are in England. I hope to-morrow may set us free from our frozen fetters.

September 17th. – The Bogie again. Thank Heaven that I have strong nerves! The superstition of these poor fellows, and the circumstantial accounts which they give, with the utmost earnestness and self-conviction, would horrify any man not accustomed to their ways. There are many versions of the matter, but the sum-total of them all is that something uncanny has been flitting round the ship all night, and that Sandie M'Donald of Peterhead and "lang" Peter Williamson of Shetland saw it, as also did Mr. Milne on the bridge – so, having three witnesses, they can make a better case of it than the second mate did. I spoke to Milne after breakfast, and told him that he should be above such nonsense, and that as an officer he ought to set the men a better example. He shook his weather-beaten head ominously, but answered with characteristic caution, "Mebbe, aye, mebbe na, Doctor," he said, "I didna ca' it a ghaist. I canna' say I preen my faith in sea-bogles an' the like, though there's a mony as claims to ha' seen a' that and waur. I'm no easy feared, but maybe your ain bluid would run a bit cauld, mun, if instead o' speerin' aboot it in daylicht ye were wi' me last night, an' seed an awfu' like shape, white an' gruesome, whiles here, whiles there, an' it greetin' and ca'ing in the darkness like a bit lambie that hae lost its mither. Ye would na' be sae ready to put it a' doon to auld wives' clavers then, I'm thinkin'." I saw it was hopeless to reason with him, so contented myself with begging him as a personal favour to call me up the next time the spectre appeared – a request to which he acceded with many ejaculations expressive of his hopes that such an opportunity might never arise.

As I had hoped, the white desert behind us has become broken by many thin streaks of water which intersect it in all directions. Our latitude to-day was 80° 52′ N., which shows that there is a strong southerly drift upon the pack. Should the wind continue favourable it will break up as rapidly as it formed. At present we can do nothing but smoke and wait and hope for the best. I am rapidly becoming a fatalist. When dealing with such uncertain factors as wind and ice a man can be nothing else. Perhaps it was the wind and sand of the Arabian deserts which gave the minds of the original followers of Mahomet their tendency to bow to kismet.

These spectral alarms have a very bad effect upon the captain. I feared that it might excite his sensitive mind, and endeavoured to conceal the absurd story from him, but unfortunately he overheard one of the men making an allusion to it, and insisted upon being informed about it. As I had expected, it brought out all his latent lunacy in an exaggerated form. I can hardly believe that this is the same man who discoursed philosophy last night with the most critical acumen and coolest judgment. He is pacing backwards and forwards upon the quarter-deck like a caged tiger, stopping now and again to throw out his hands with a yearning gesture, and stare impatiently out over the ice. He keeps up a continual mutter to himself, and once he called out, "But a little time, love – but a little time!" Poor fellow, it is sad to see a gallant seaman and accomplished gentleman reduced to such a pass, and to think that imagination and delusion can cow a mind to which real danger was but the salt of life. Was ever a man in such a position as I, between a demented captain and a ghost-seeing mate? I sometimes think I am the only really sane man aboard the vessel – except perhaps the second engineer, who is a kind of ruminant, and would care nothing for all the fiends in the Red Sea so long as they would leave him alone and not disarrange his tools.

The ice is still opening rapidly, and there is every probability of our being able to make a start to-morrow morning. They will think I am inventing when I tell them at home all the strange things that have befallen me.

12 P.M. – I have been a good deal startled, though I feel steadier now, thanks to a stiff glass of brandy. I am hardly myself yet, however, as, this handwriting will testify. The fact is, that I have gone through a very strange experience, and am beginning to doubt whether I was justified in branding everyone on board as madmen because they professed to have seen things which did not seem reasonable to my understanding. Pshaw! I am a fool to let such a trifle unnerve me; and yet, coming as it does after all these alarms, it has an additional significance, for I cannot doubt either Mr. Manson's story or that of the mate, now that I have experienced that which I used formerly to scoff at.

After all it was nothing very alarming – a mere sound, and that was all. I cannot expect that anyone reading this, if anyone ever should read it, will sympathise with my feelings, or realize the effect which it produced upon me at the time. Supper was over, and I had gone on deck to have a quiet pipe before turning in. The night was very dark – so dark that, standing under the quarter-boat, I was unable to see the officer upon the bridge. I think I have already mentioned the extraordinary silence which prevails in these frozen seas. In other parts of the world, be they ever so barren, there is some slight vibration of the air – some faint hum, be it from the distant haunts of men, or from the leaves of the trees, or the wings of the birds, or even the faint rustle of the grass that covers the ground. One may not actively perceive the sound, and yet if it were withdrawn it would be missed. It is only here in these Arctic seas that stark, unfathomable stillness obtrudes itself upon you all in its gruesome reality. You find your tympanum straining to catch some little murmur, and dwelling eagerly upon every accidental sound within the vessel. In this state I was leaning against the bulwarks when there arose from the ice almost directly underneath me a cry, sharp and shrill, upon the silent air of the night, beginning, as it seemed to me, at a note such as prima donna never reached, and mounting from that ever higher and higher until it culminated in a long wail of agony, which might have been the last cry of a lost soul. The ghastly scream is still ringing in my ears. Grief, unutterable grief, seemed to be expressed in it, and a great longing, and yet through it all there was an occasional wild note of exultation. It shrilled out from close beside me, and yet as I glared into the darkness I could discern nothing. I waited some little time, but without hearing any repetition of the sound, so I came below, more shaken than I have ever been in my life before. As I came down the companion I met Mr. Milne coming up to relieve the watch. "Weel, Doctor," he said, "maybe that's auld wives' clavers tae? Did ye no hear it skirling? Maybe that's a supersteetion? What d'ye think o't noo?" I was obliged to apologise to the honest fellow, and acknowledge that I was as puzzled by it as he was. Perhaps to-morrow things may look different. At present I dare hardly write all that I think. Reading it again in

days to come, when I have shaken off all these associations, I should despise myself for having been so weak.

September 18th. – Passed a restless and uneasy night, still haunted by that strange sound. The captain does not look as if he had had much repose either, for his face is haggard and his eyes bloodshot. I have not told him of my adventure of last night, nor shall I. He is already restless and excited, standing up, sitting down, and apparently utterly unable to keep still.

A fine lead appeared in the pack this morning, as I had expected, and we were able to cast off our ice-anchor, and steam about twelve miles in a west-sou'-westerly direction. We were then brought to a halt by a great floe as massive as any which we have left behind us. It bars our progress completely, so we can do nothing but anchor again and wait until it breaks up, which it will probably do within twenty-four hours, if the wind holds. Several bladder-nosed seals were seen swimming in the water, and one was shot, an immense creature more than eleven feet long. They are fierce, pugnacious animals, and are said to be more than a match for a bear. Fortunately they are slow and clumsy in their movements, so that there is little danger in attacking them upon the ice.

The captain evidently does not think we have seen the last of our troubles, though why he should take a gloomy view of the situation is more than I can fathom, since everyone else on board considers that we have had a miraculous escape, and are sure now to reach the open sea.

"I suppose you think it's all right now, Doctor?" he said, as we sat together after dinner.

"I hope so," I answered.

"We mustn't be too sure – and yet no doubt you are right. We'll all be in the arms of our own true loves before long, lad, won't we? But we mustn't be too sure – we mustn't be too sure."

He sat silent a little, swinging his leg thoughtfully backwards and forwards. "Look here," he continued; "it's a dangerous place this, even at its best – a treacherous, dangerous place. I have known men cut off very suddenly in a land like this. A slip would do it sometimes – a single slip, and down you go through a crack, and only a bubble on the green water to show where it was that you sank. It's a queer thing," he continued with a nervous laugh, "but all the years I've been in this country I never once thought of making a will – not that I have anything to leave in particular, but still when a man is exposed to danger he should have everything arranged and ready – don't you think so?"

"Certainly," I answered, wondering what on earth he was driving at.

"He feels better for knowing it's all settled," he went on. "Now if anything should ever befall me, I hope that you will look after things for me. There is very little in the cabin, but such as it is I should like it to be sold, and the money divided in the same proportion as the oil-money among the crew. The chronometer I wish you to keep yourself as some slight remembrance of our voyage. Of course all this is a mere precaution, but I thought I would take this opportunity of speaking to you about it. I suppose I might rely upon you if there were any necessity?"

"Most assuredly," I answered, "and since you are taking this step, I may as well – "

"You! you!" he interrupted. "You're all right. What the devil is the matter with you? There, I didn't mean to be peppery, but I don't like to hear a young fellow, that has hardly began life, speculating about death. Go up on deck and get some fresh air into your lungs instead of talking nonsense in the cabin, and encouraging me to do the same."

The more I think of this conversation of ours the less do I like it. Why should the man be settling his affairs at the very time when we seem to be emerging from all danger? There must be some method in his madness. Can it be that he contemplates suicide? I remember that upon one occasion he spoke in a deeply reverent manner of the heinousness of the crime of self-destruction. I shall keep my eye upon him, however, and though I cannot obtrude upon the privacy of his cabin, I shall at least make a point of remaining on deck as long as he stays up.

Mr. Milne pooh-poohs my fears, and says it is only the "skipper's little way." He himself takes a very rosy view of the situation. According to him we shall be out of the ice by the day after to-morrow, pass Jan Mayen two days after that, and sight Shetland in little more than a week. I hope he may not be too sanguine. His opinion may be fairly balanced against the gloomy precautions of the captain, for he is an old and experienced seaman, and weighs his words well before uttering them.

* * *

The long-impending catastrophe has come at last. I hardly know what to write about it. The captain is gone. He may come back to us again alive, but I fear me – I fear me. It is now seven o'clock of the morning of the 19th of September. I have spent the whole night traversing the great ice-floe in front of us with a party of seamen in the hope of coming upon some trace of him, but in vain. I shall try to give some account of the circumstances which attended upon his disappearance. Should anyone ever chance to read the words which I put down, I trust they will remember that I do not write from conjecture or from hearsay, but that I, a sane and educated man, am describing accurately what actually occurred before my very eyes. My inferences are my own, but I shall be answerable for the facts.

The captain remained in excellent spirits after the conversation which I have recorded. He appeared to be nervous and impatient, however, frequently changing his position, and moving his limbs in an aimless choreic way which is characteristic of him at times. In a quarter of an hour he went upon deck seven times, only to descend after a few hurried paces. I followed him each time, for there was something about his face which confirmed my resolution of not letting him out of my sight. He seemed to observe the effect which his movements had produced, for he endeavoured by an overdone hilarity, laughing boisterously at the very smallest of jokes, to quiet my apprehensions.

After supper he went on to the poop once more, and I with him. The night was dark and very still, save for the melancholy soughing of the wind among the spars. A thick cloud was coming up from the north-west, and the ragged tentacles which it threw out in front of it were drifting across the face of the moon, which only shone now and again through a rift in the wrack. The captain paced rapidly backwards and forwards, and then seeing me still dogging him, he came across and hinted that he thought I should be better below – which, I need hardly say, had the effect of strengthening my resolution to remain on deck.

I think he forgot about my presence after this, for he stood silently leaning over the taffrail and peering out across the great desert of snow, part of which lay in shadow, while part glittered mistily in the moonlight. Several times I could see by his movements that he was referring to his watch, and once he muttered a short sentence, of which I could only catch the one word "ready." I confess to having felt an eerie feeling creeping over me as I watched the loom of his tall figure through the darkness, and noted how completely he fulfilled the idea of a man who is keeping a tryst. A tryst with whom? Some vague perception began to dawn upon me as I pieced one fact with another, but I was utterly unprepared for the sequel.

By the sudden intensity of his attitude I felt that he saw something. I crept up behind him. He was staring with an eager questioning gaze at what seemed to be a wreath of mist, blown swiftly in a line with the ship. It was a dim nebulous body, devoid of shape, sometimes more, sometimes less apparent, as the light fell on it. The moon was dimmed in its brilliancy at the moment by a canopy of thinnest cloud, like the coating of an anemone.

"Coming, lass, coming," cried the skipper, in a voice of unfathomable tenderness and compassion, like one who soothes a beloved one by some favour long looked for, and as pleasant to bestow as to receive.

What followed happened in an instant. I had no power to interfere. He gave one spring to the top of the bulwarks, and another which took him on to the ice, almost to the feet of the pale misty figure. He held out his hands as if to clasp it, and so ran into the darkness with outstretched arms and loving words. I still stood rigid and motionless, straining my eyes after his retreating form, until his voice died away in the distance. I never thought to see him again, but at that moment the moon shone out brilliantly through a chink in the cloudy heaven, and illuminated the great field of ice. Then I saw his dark figure already a very long way off, running with prodigious speed across the frozen plain. That was the last glimpse which we caught of him – perhaps the last we ever shall. A party was organized to follow him, and I accompanied them, but the men's hearts were not in the work, and nothing was found. Another will be formed within a few hours. I can hardly believe I have not been dreaming, or suffering from some hideous nightmare, as I write these things down.

7.30 P.M. – Just returned dead beat and utterly tired out from a second unsuccessful search for the captain. The floe is of enormous extent, for though we have traversed at least twenty miles of its surface, there has been no sign of its coming to an end. The frost has been so severe of late that the overlying snow is frozen as hard as granite, otherwise we might have had the footsteps to guide us. The crew are anxious that we should cast off and steam round the floe and so to the southward, for the ice has opened up during the night, and the sea is visible upon the horizon. They argue that Captain Craigie is certainly dead, and that we are all risking our lives to no purpose by remaining when we have an opportunity of escape. Mr. Milne and I have had the greatest difficulty in persuading them to wait until to-morrow night, and have been compelled to promise that we will not under any circumstances delay our departure longer than that. We propose therefore to take a few hours' sleep, and then to start upon a final search.

September 20th, evening. – I crossed the ice this morning with a party of men exploring the southern part of the floe, while Mr. Milne went off in a northerly direction. We pushed on for ten or twelve miles without seeing a trace of any living thing except a single bird, which fluttered a great way over our heads, and which by its flight I should judge to have been a falcon. The southern extremity of the ice-field tapered away into a long narrow spit which projected out into the sea. When we came to the base of this promontory, the men halted, but I begged them to continue to the extreme end of it, that we might have the satisfaction of knowing that no possible chance had been neglected.

We had hardly gone a hundred yards before M'Donald of Peterhead cried out that he saw something in front of us, and began to run. We all got a glimpse of it and ran too. At first it was only a vague darkness against the white ice, but as we raced along together it took the shape of a man, and eventually of the man of whom we were in search. He was lying face downwards upon a frozen bank. Many little crystals of ice and feathers of snow had drifted on to him as he lay, and sparkled upon his dark seaman's jacket. As we came up some wandering puff of wind caught these tiny flakes in its vortex, and they whirled up into the air, partially descended again, and then, caught once more in the current, sped rapidly away in the direction of the sea. To my eyes it seemed but a snow-drift, but many of my companions averred that it started up in the shape of a woman, stooped over the corpse and kissed it, and then hurried away across the floe. I have learned never to ridicule any man's opinion, however strange it may seem. Sure it is that Captain Nicholas Craigie had met with no painful end, for there was a bright smile upon his blue pinched features, and his hands were still outstretched as though grasping at the strange visitor which had summoned him away into the dim world that lies beyond the grave.

We buried him the same afternoon with the ship's ensign around him, and a thirty-two pound shot at his feet. I read the burial service, while the rough sailors wept like children, for there were many who owed much to his kind heart, and who showed now the affection which his strange ways had repelled during his lifetime. He went off the grating with a dull, sullen splash, and as I looked into the green water I saw him go down, down, down until he was but a little flickering patch of white hanging upon the outskirts of eternal darkness. Then even that faded away, and he was gone. There he shall lie, with his secret and his sorrows and his mystery all still buried in his breast, until that great day when the sea shall give up its dead, and Nicholas Craigie come out from among the ice with the smile upon his face, and his stiffened arms outstretched in greeting. I pray that his lot may be a happier one in that life than it has been in this.

I shall not continue my journal. Our road to home lies plain and clear before us, and the great ice-field will soon be but a remembrance of the past. It will be some time before I get over the shock produced by recent events. When I began this record of our voyage I little thought of how I should be compelled to finish it. I am writing these final words in the lonely cabin, still starting at times and fancying I hear the quick nervous step of the dead man upon the deck above me. I entered his cabin to-night, as was my duty, to make a list of his effects in order that they might be entered in the official log. All was as it had been upon my previous visit, save that the picture which I have described as having hung at the end of his bed had been cut out of its frame, as with a knife, and was gone. With this last link in a strange chain of evidence I close my diary of the voyage of the *Pole-Star*.

[NOTE by Dr. John M'Alister Ray, senior. – I have read over the strange events connected with the death of the captain of the *Pole-Star*, as narrated in the journal of my son. That everything occurred exactly as he describes it I have the fullest confidence, and, indeed the most positive certainty, for I know him to be a strong-nerved and unimaginative man, with the strictest regard for veracity. Still, the story is, on the face of it, so vague and so improbable, that I was long opposed to its publication. Within the last few days, however, I have had independent testimony upon the subject which throws a new light upon it. I had run down to Edinburgh to attend a meeting of the British Medical Association, when I chanced to come across Dr. P ———, an old college chum of mine, now practicing at Saltash, in Devonshire. Upon my telling him of this experience of my son's, he declared to me that he was familiar with the man, and proceeded, to my no small surprise, to give me a description of him, which tallied remarkably well with that given in the journal, except that he depicted him as a younger man. According to his account, he had been engaged to a young lady of singular beauty residing upon the Cornish coast. During his absence at sea his betrothed had died under circumstances of peculiar horror.]

The Adventure of Black Peter

EDITORS' NOTE   Though the Sherlock Holmes tales are filled with ships and sailors of all descriptions, none so clearly shows the mark of Conan Doyle's experiences aboard the *Hope* as "The Adventure of Black Peter," in which the captain of a whaler is found "pinned like a beetle on a card" by a steel harpoon thrust through his chest. Some of Conan Doyle's biographers have reported that the story's appearance in *The Strand Magazine*, March 1904, drew an angry letter from John Gray, suggesting the captain of the *Hope* saw something of himself in the villainous Peter Carey. This seems unlikely, not only because Carey hailed from the rival port Dundee, but also because Captain Gray had died at home in Peterhead a dozen years before the story's publication.

I HAVE NEVER known my friend to be in better form, both mental and physical, than in the year '95. His increasing fame had brought with it an immense practice, and I should be guilty of an indiscretion if I were even to hint at the identity of some of the illustrious clients who crossed our humble threshold in Baker Street. Holmes, however, like all great artists, lived for his art's sake, and, save in the case of the Duke of Holdernesse, I have seldom known him claim any large reward for his inestimable services. So unworldly was he – or so capricious – that he frequently refused his help to the powerful and wealthy where the problem made no appeal to his sympathies, while he would devote weeks of most intense application to the affairs of some humble client whose case presented those strange and dramatic qualities which appealed to his imagination and challenged his ingenuity.

In this memorable year '95, a curious and incongruous succession of cases had engaged his attention, ranging from his famous investigation of the sudden death of Cardinal Tosca – an inquiry which was carried out by him at the express desire of His Holiness the Pope – down to his arrest of Wilson, the notorious canary-trainer, which removed a plague-spot from the East End of London. Close on the heels of these two famous cases came the tragedy of Woodman's Lee, and the very obscure circumstances which surrounded the death of Captain Peter Carey.

No record of the doings of Mr. Sherlock Holmes would be complete which did not include some account of this very unusual affair.

During the first week of July, my friend had been absent so often and so long from our lodgings that I knew he had something on hand. The fact that several rough-looking men called during that time and inquired for Captain Basil made me understand that Holmes was working somewhere under one of the numerous disguises and names with which he concealed his own formidable identity. He had at least five small refuges in different parts of London in which he was able to change his personality. He said nothing of his business to me, and it was not my habit to force a confidence. The first positive sign which he gave me of the direction which his investigation was taking was an extraordinary one. He had gone out before breakfast, and I had sat down to mine when he strode into the room, his hat upon his head and a huge barbed-headed spear tucked like an umbrella under his arm.

"Good gracious, Holmes!" I cried. "You don't mean to say that you have been walking about London with that thing?"

"I drove to the butcher's and back."

"The butcher's?"

"And I return with an excellent appetite. There can be no question, my dear Watson, of the value of exercise before breakfast. But I am prepared to bet that you will not guess the form that my exercise has taken."

"I will not attempt it."

He chuckled as he poured out the coffee.

"If you could have looked into Allardyce's back shop you would have seen a dead pig swung from a hook in the ceiling, and a gentleman in his shirt sleeves furiously stabbing at it with this weapon. I was that energetic person, and I have satisfied myself that by no exertion of my strength can I transfix the pig with a single blow. Perhaps you would care to try?"

"Not for worlds. But why were you doing this?"

"Because it seemed to me to have an indirect bearing upon the mystery of Woodman's Lee. Ah, Hopkins, I got your wire last night, and I have been expecting you. Come and join us."

Our visitor was an exceedingly alert man, thirty years of age, dressed in a quiet tweed suit, but retaining the erect bearing of one who was accustomed to official uniform. I recognized him at once as Stanley Hopkins, a young police inspector for whose future Holmes had high hopes, while he in turn professed the admiration and respect of a pupil for the scientific methods of the famous amateur. Hopkins's brow was clouded and he sat down with an air of deep dejection.

"No, thank you, sir. I breakfasted before I came round. I spent the night in town, for I came up yesterday to report."

"And what had you to report?"

"Failure, sir – absolute failure."

"You have made no progress?"

"None."

"Dear me! I must have a look at the matter."

"I wish to Heavens that you would, Mr. Holmes. It's my first big chance, and I am at my wit's end. For goodness' sake, come down and lend me a hand."

"Well, well, it just happens that I have already read all the available evidence, including the report of the inquest, with some care. By the way, what do you make of that tobacco-pouch found on the scene of the crime? Is there no clue there?"

Hopkins looked surprised.

"It was the man's own pouch, sir. His initials were inside it. And it was of sealskin – and he an old sealer."

"But he had no pipe."

"No, sir, we could find no pipe; indeed, he smoked very little. And yet he might have kept some tobacco for his friends."

"No doubt. I only mention it because if I had been handling the case I should have been inclined to make that the starting-point of my investigation. However, my friend Dr. Watson knows nothing of this matter, and I should be none the worse for hearing the sequence of events once more. Just give us some short sketches of the essentials."

Stanley Hopkins drew a slip of paper from his pocket.

"I have a few dates here which will give you the career of the dead man, Captain Peter Carey. He was born in '45 – fifty years of age. He was a most daring and successful seal and whale fisher. In 1883 he commanded the steam sealer *Sea Unicorn*, of Dundee. He had then had several successful voyages in succession, and in the following year, 1884, he retired. After that he travelled for some years, and finally he bought a small place called Woodman's Lee, near Forest Row, in Sussex. There he has lived for six years, and there he died just a week ago to-day.

"There were some most singular points about the man. In ordinary life, he was a strict Puritan – a silent, gloomy fellow. His household consisted of his wife, his daughter, aged twenty, and two female servants. These last were continually changing, for it was never a very cheery situation, and sometimes it became past all bearing. The man was an intermittent drunkard, and when he had the fit on him he was a perfect fiend. He has been known to drive his wife and daughter out of doors in the middle of the night, and flog them through the park until the whole village outside the gates was aroused by their screams.

"He was summoned once for a savage assault upon the old vicar, who had called upon him to remonstrate with him upon his conduct. In short, Mr. Holmes, you would go far before you found a more dangerous man than Peter Carey, and I have heard that he bore the same character when he commanded his ship. He was known in the trade as Black Peter, and the name was given him, not only on account of his swarthy features and the colour of his huge beard, but for the humours which were the terror of all around him. I need not say that he was loathed and avoided by every one of his neighbours, and that I have not heard one single word of sorrow about his terrible end.

"You must have read in the account of the inquest about the man's cabin, Mr. Holmes; but perhaps your friend here has not heard of it. He had built himself a wooden outhouse – he always called it 'the cabin' – a few hundred yards from his house, and it was here that he slept every night. It was a little, single-roomed hut, sixteen feet by ten. He kept the key in his pocket, made his own bed, cleaned it himself, and allowed no other foot to cross the threshold. There are small windows on each side, which were covered by curtains, and never opened. One of these windows was turned towards the high-road, and when the light burned in it at night the folk used to point it out to each other, and wonder what Black Peter was doing in there. That's the window, Mr. Holmes, which gave us one of the few bits of positive evidence that came out at the inquest.

"You remember that a stonemason, named Slater, walking from Forest Row about one o'clock in the morning – two days before the murder – stopped as he passed the grounds and looked at the square of light still shining among the trees. He swears that the shadow of a man's head turned sideways was clearly visible on the blind, and that this shadow was certainly not that of Peter Carey, whom he knew well. It was that of a bearded man, but the beard was short, and bristled forward in a way very different from that of the captain. So he says, but he had been two hours in the public-house, and it is some distance from the road to the window. Besides, this refers to the Monday, and the crime was done upon the Wednesday.

"On the Tuesday Peter Carey was in one of his blackest moods, flushed with drink and as savage as a dangerous wild beast. He roamed about the house, and the women ran for it when they heard him coming. Late in the evening he went down to his own hut. About two o'clock the following morning his daughter, who slept with her window open, heard a most fearful yell from that direction, but it was no unusual thing for him to bawl and shout when he was in drink, so no notice was taken. On rising at seven one of the maids noticed that the door of the hut was open, but so great was the terror which the man caused that it was midday before anyone would venture down to see what had become of him. Peeping into the open door, they saw a sight which sent them flying with white faces into the village. Within an hour I was on the spot and had taken over the case.

"Well, I have fairly steady nerves, as you know, Mr. Holmes, but I give you my word that I got a shake when I put my head into that little house. It was droning like a harmonium with the flies and bluebottles, and the floor and walls were like a slaughter-house. He had called it a cabin, and a cabin it was, sure enough, for you would have thought that you were in a ship. There was a bunk at one end, a sea-chest, maps and charts, a

picture of the *Sea Unicorn*, a line of log-books on a shelf, all exactly as one would expect to find it in a captain's room. And there in the middle of it was the man himself, his face twisted like a lost soul in torment, and his great brindled beard stuck upward in his agony. Right through his broad breast a steel harpoon had been driven, and it had sunk deep into the wood of the wall behind him. He was pinned like a beetle on a card. Of course, he was quite dead, and had been so from the instant that he had uttered that last yell of agony.

"I know your methods, sir, and I applied them. Before I permitted anything to be moved I examined most carefully the ground outside, and also the floor of the room. There were no footmarks."

"Meaning that you saw none?"

"I assure you, sir, that there were none."

"My good Hopkins, I have investigated many crimes, but I have never yet seen one which was committed by a flying creature. As long as the criminal remains upon two legs so long must there be some indentation, some abrasion, some trifling displacement which can be detected by the scientific searcher. It is incredible that this blood-bespattered room contained no trace which could have aided us. I understand, however, from the inquest that there were some objects which you failed to overlook?"

The young inspector winced at my companion's ironical comments.

"I was a fool not to call you in at the time, Mr. Holmes. However, that's past praying for now. Yes, there were several objects in the room which called for special attention.

"One was the harpoon with which the deed was committed. It had been snatched down from a rack on the wall. Two others remained there, and there was a vacant place for the third. On the stock was engraved 'S.S. *Sea Unicorn*, Dundee.' This seemed to establish that the crime had been done in a moment of fury, and that the murderer had seized the first weapon which came in his way. The fact that the crime was committed at two in the morning, and yet Peter Carey was fully dressed, suggested that he had an appointment with the murderer, which is borne out by the fact that a bottle of rum and two dirty glasses stood upon the table."

"Yes," said Holmes; "I think that both inferences are permissible. Was there any other spirit but rum in the room?"

"Yes, there was a tantalus containing brandy and whisky on the sea-chest. It is of no importance to us, however, since the decanters were full and it had therefore not been used."

"For all that, its presence had some significance," said Holmes. "However, let us hear some more about the objects which do seem to you to bear upon the case."

"There was this tobacco-pouch upon the table."

"What part of the table?"

"It lay in the middle. It was of coarse sealskin – the straight-haired skin, with a leather thong to bind it. Inside was 'P. C.' on the flap. There was half an ounce of strong ship's tobacco in it."

"Excellent! What more?"

Stanley Hopkins drew from his pocket a drab-covered notebook. The outside was rough and worn, the leaves were discoloured. On the first page were written the initials "J. H. N." and the date "1883." Holmes laid it on the table and examined it in his minute way, while Hopkins and I gazed over each shoulder. On the second page were printed the letters "C. P. R." and then came several sheets of numbers. Another heading was

"Argentine," another "Costa Rica," and another "San Paulo," each with pages of signs and figures after it.

"What do you make of these?" asked Holmes.

"They appear to be lists of Stock Exchange securities. I thought that 'J. H. N.' were the initials of a broker, and that 'C. P. R.' may have been his client."

"Try Canadian Pacific Railway," said Holmes.

Stanley Hopkins swore between his teeth, and struck his thigh with his clenched hand.

"What a fool I have been!" he cried. "Of course it is as you say. Then 'J. H. N.' are the only initials we have to solve. I have already examined the old Stock Exchange lists, and I can find no one in 1883 either in the House or among the outside brokers whose initials correspond with these. Yet I feel that the clue is the most important that I hold. You will admit, Mr. Holmes, that there is a possibility that these initials are those of the second person who was present – in other words, of the murderer. I would also urge that the introduction into the case of a document relating to large masses of valuable securities gives us for the first time some indication of a motive for the crime."

Sherlock Holmes's face showed that he was thoroughly taken back by this new development.

"I must admit both your points," said he. "I confess that this notebook, which did not appear at the inquest, modifies any views which I may have formed. I had come to a theory of the crime in which I can find no place for this. Have you endeavoured to trace any of the securities here mentioned?"

"Inquiries are now being made at the offices, but I fear that the complete register of the stock-holders of these

South American concerns is in South America, and that some weeks must elapse before we can trace the shares."

Holmes had been examining the cover of the notebook with his magnifying lens.

"Surely there is some discolouration here," said he.

"Yes, sir, it is a blood-stain. I told you that I picked the book off the floor."

"Was the blood-stain above or below?"

"On the side next the boards."

"Which proves, of course, that the book was dropped after the crime was committed."

"Exactly, Mr. Holmes. I appreciated that point, and I conjectured that it was dropped by the murderer in his hurried flight. It lay near the door."

"I suppose that none of these securities have been found among the property of the dead man?"

"No, sir."

"Have you any reason to suspect robbery?"

"No, sir. Nothing seemed to have been touched."

"Dear me, it is certainly a very interesting case. Then there was a knife, was there not?"

"A sheath-knife, still in its sheath. It lay at the feet of the dead man. Mrs. Carey has identified it as being her husband's property."

Holmes was lost in thought for some time.

"Well," said he, at last, "I suppose I shall have to come out and have a look at it."

Stanley Hopkins gave a cry of joy.

"Thank you, sir. That will indeed be a weight off my mind."

Holmes shook his finger at the inspector.

"It would have been an easier task a week ago," said he. "But even now

my visit may not be entirely fruitless. Watson, if you can spare the time, I should be very glad of your company. If you will call a four-wheeler, Hopkins, we shall be ready to start for Forest Row in a quarter of an hour."

Alighting at the small wayside station, we drove for some miles through the remains of widespread woods, which were once part of that great forest which for so long held the Saxon invaders at bay – the impenetrable "weald," for sixty years the bulwark of Britain. Vast sections of it have been cleared, for this is the seat of the first ironworks of the country, and the trees have been felled to smelt the ore. Now the richer fields of the North have absorbed the trade, and nothing save these ravaged groves and great scars in the earth show the work of the past. Here in a clearing upon the green slope of a hill stood a long, low stone house, approached by a curving drive running through the fields. Nearer the road, and surrounded on three sides by bushes, was a small outhouse, one window and the door facing in our direction. It was the scene of the murder.

Stanley Hopkins led us first to the house, where he introduced us to a haggard, grey-haired woman, the widow of the murdered man, whose gaunt and deep-lined face, with the furtive look of terror in the depths of her red-rimmed eyes, told of the years of hardship and ill-usage which she had endured. With her was her daughter, a pale, fair-haired girl, whose eyes blazed defiantly at us as she told us that she was glad that her father was dead, and that she blessed the hand which had struck him down. It was a terrible household that Black Peter Carey had made for himself, and it was with a sense of relief that we found ourselves in the sunlight again and making our way along a path which had been worn across the fields by the feet of the dead man.

The outhouse was the simplest of dwellings, wooden-walled, shingle-roofed, one window beside the door, and one on the farther side. Stanley Hopkins drew the key from his pocket, and had stooped to the lock, when he paused with a look of attention and surprise upon his face.

"Someone has been tampering with it," he said.

There could be no doubt of the fact. The woodwork was cut, and the scratches showed white through the paint, as if they had been that instant done. Holmes had been examining the window.

"Someone has tried to force this also. Whoever it was has failed to make his way in. He must have been a very poor burglar."

"This is a most extraordinary thing," said the inspector; "I could swear that these marks were not here yesterday evening."

"Some curious person from the village, perhaps," I suggested.

"Very unlikely. Few of them would dare to set foot in the grounds, far less try to force their way into the cabin. What do you think of it, Mr. Holmes?"

"I think that fortune is very kind to us."

"You mean that the person will come again?"

"It is very probable. He came expecting to find the door open. He tried to get in with the blade of a very small penknife. He could not manage it. What would he do?"

"Come again next night with a more useful tool."

"So I should say. It will be our fault if we are not there to receive him. Meanwhile, let me see the inside of the cabin."

The traces of the tragedy had been removed, but the furniture within the little room still stood as it had been on the night of the crime. For two hours, with most intense concentration, Holmes examined every object in turn, but his face showed that his quest was not a successful one. Once only he paused in his patient investigation.

"Have you taken anything off this shelf, Hopkins?"

"No; I have moved nothing."

"Something has been taken. There is less dust in this corner of the shelf than elsewhere. It may have been a book lying on its side. It may have been a box. Well, well, I can do nothing more. Let us walk in these beautiful woods, Watson, and give a few hours to the birds and the flowers. We shall meet you here later, Hopkins, and see if we can come to close quarters with the gentleman who has paid his visit in the night."

It was past eleven o'clock when we formed our little ambuscade. Hopkins was for leaving the door of the hut open, but Holmes was of the opinion that this would rouse the suspicions of the stranger. The lock was a perfectly simple one, and only a strong blade was needed to push it back. Holmes also suggested that we should wait, not inside the hut, but outside it among the bushes which grew round the farther window. In this way we should be able to watch our man if he struck a light, and see what his object was in this stealthy nocturnal visit.

It was a long and melancholy vigil, and yet it brought with it something of the thrill which the hunter feels when he lies beside the water-pool and waits for the coming of the thirsty beast of prey. What savage creature was it which might steal upon us out of the darkness? Was it a fierce tiger of crime, which could only be taken fighting hard with flashing fang and claw, or would it prove to be some skulking jackal, dangerous only to the weak and unguarded? In absolute silence we crouched amongst the bushes, waiting for whatever might come. At first the steps of a few belated villagers, or the sound of voices from the village, lightened our vigil; but one by one these interruptions died away, and an absolute stillness fell upon us, save for the chimes of the distant church, which told us of the progress of the night, and for the rustle and whisper of a fine rain falling amid the foliage which roofed us in.

Half-past two had chimed, and it was the darkest hour which precedes the dawn, when we all started as a low but sharp click came from the direction of the gate. Someone had entered the drive. Again there was a long silence, and I had begun to fear that it was a false alarm, when a stealthy step was heard upon the other side of the hut, and a moment later a metallic scraping and clinking. The man was trying to force the lock! This time his skill was greater or his tool was better, for there was a sudden snap and the creak of the hinges. Then a match was struck, and next instant the steady light from a candle filled the interior of the hut. Through the gauze curtain our eyes were all riveted upon the scene within.

The nocturnal visitor was a young man, frail and thin, with a black moustache which intensified the deadly pallor of his face. He could not have been much above twenty years of age. I have never seen any human being who appeared to be in such a pitiable fright, for his teeth were visibly chattering, and he was shaking in every limb. He was dressed like a gentleman, in Norfolk jacket and knickerbockers, with a cloth cap upon his head. We watched him staring round with frightened eyes. Then he laid the candle-end upon the table

and disappeared from our view into one of the corners. He returned with a large book, one of the log-books which formed a line upon the shelves. Leaning on the table, he rapidly turned over the leaves of this volume until he came to the entry which he sought. Then, with an angry gesture of his clenched hand, he closed the book, replaced it in the corner, and put out the light. He had hardly turned to leave the hut when Hopkins's hand was on the fellow's collar, and I heard his loud gasp of terror as he understood that he was taken. The candle was relit, and there was our wretched captive shivering and cowering in the grasp of the detective. He sank down upon the sea-chest, and looked helplessly from one of us to the other.

"Now, my fine fellow," said Stanley Hopkins, "who are you, and what do you want here?"

The man pulled himself together and faced us with an effort at self-composure.

"You are detectives, I suppose?" said he. "You imagine I am connected with the death of Captain Peter Carey. I assure you that I am innocent."

"We'll see about that," said Hopkins. "First of all, what is your name?"

"It is John Hopley Neligan."

I saw Holmes and Hopkins exchange a quick glance.

"What are you doing here?"

"Can I speak confidentially?"

"No, certainly not."

"Why should I tell you?"

"If you have no answer it may go badly with you at the trial."

The young man winced.

"Well, I will tell you," he said. "Why should I not? And yet I hate to think of this old scandal gaining a new lease of life. Did you ever hear of Dawson and Neligan?"

I could see from Hopkins's face that

he never had; but Holmes was keenly interested.

"You mean the West Country bankers," said he. "They failed for a million, ruined half the county families of Cornwall, and Neligan disappeared."

"Exactly. Neligan was my father."

At last we were getting something positive, and yet it seemed a long gap between an absconding banker and Captain Peter Carey pinned against the wall with one of his own harpoons. We all listened intently to the young man's words.

"It was my father who was really concerned. Dawson had retired. I was only ten years of age at the time, but I was old enough to feel the shame and horror of it all. It has always been said that my father stole all the securities and fled. It is not true. It was his belief that if he were given time in which to realize them, all would be well, and every creditor paid in full. He started in his little yacht for Norway just before the warrant was issued for his arrest. I can remember that last night when he bade farewell to my mother. He left us a list of the securities he was taking, and he swore that he would come back with his honour cleared, and that none who had trusted him would suffer. Well, no word was ever heard from him again. Both the yacht and he vanished utterly. We believed, my mother and I, that he and it, with the securities that he had taken with him, were at the bottom of the sea. We had a faithful friend, however, who is a business man, and it was he who discovered some time ago that some of the securities which my father had with him had reappeared on the London market. You can imagine our amazement. I spent months in trying to trace them, and at last, after many doubtings and difficulties, I discovered that the original seller had been Captain Peter Carey, the owner of this hut.

"Naturally I made some inquiries about the man. I found that he had been in command of a whaler which was due to return from the Arctic seas at the very time when my father was crossing to Norway. The autumn of that year was a stormy one, and there was a long succession of southerly gales. My father's yacht may well have been blown to the north, and there met by Captain Peter Carey's ship. If that were so, what had become of my father? In any case, if I could prove from Peter Carey's evidence how these securities came on the market, it would be a proof that my father had not sold them, and that he had no view to personal profit when he took them.

"I came down to Sussex with the intention of seeing the captain, but it was at this moment that his terrible death occurred. I read at the inquest a description of his cabin, in which it stated that the old log-books of his vessel were preserved in it. It struck me that if I could see what occurred in the month of August, 1883, on board the *Sea Unicorn*, I might settle the mystery of my father's fate. I tried last night to get at these log-books, but was unable to open the door. To-night I tried again, and succeeded; but I find that the pages which deal with that month have been torn from the book. It was at that moment I found myself a prisoner in your hands."

"Is that all?" asked Hopkins.

"Yes, that is all." His eyes shifted as he said it.

"You have nothing else to tell us?"

He hesitated.

"No; there is nothing."

"You have not been here before last night?"

"No."

"Then how do you account for

that?" cried Hopkins, as he held up the damning notebook, with the initials of our prisoner on the first leaf, and the blood-stain on the cover.

The wretched man collapsed. He sank his face in his hands, and trembled all over. "Where did you get it?" he groaned. "I did not know. I thought I had lost it at the hotel."

"That is enough," said Hopkins sternly. "Whatever else you have to say you must say in court. You will walk down with me now to the police-station. Well, Mr. Holmes, I am very much obliged to you and to your friend for coming down to help me. As it turns out your presence was unnecessary, and I would have brought the case to this successful issue without you; but none the less I am very grateful. Rooms have been reserved for you at the Brambletye Hotel, so we can all walk down to the village together."

"Well, Watson, what do you think of it?" asked Holmes as we travelled back next morning.

"I can see that you are not satisfied."

"Oh, yes, my dear Watson, I am perfectly satisfied. At the same time Stanley Hopkins's methods do not commend themselves to me. I am disappointed in Stanley Hopkins. I had hoped for better things from him. One should always look for a possible alternative and provide against it. It is the first rule of criminal investigation."

"What, then, is the alternative?"

"The line of investigation which I have myself been pursuing. It may give us nothing. I cannot tell. But at least I shall follow it to the end."

Several letters were waiting for Holmes at Baker Street. He snatched one of them up, opened it, and burst out into a triumphant chuckle of laughter.

"Excellent, Watson. The alternative develops. Have you telegraph forms? Just write a couple of messages for me: 'Sumner, Shipping Agent, Ratcliff Highway. Send three men on, to arrive ten to-morrow morning. – Basil.' That's my name in those parts. The other is 'Inspector Stanley Hopkins, 46, Lord Street, Brixton. Come breakfast to-morrow at nine-thirty. Important. Wire if unable to come. – Sherlock Holmes.' There, Watson, this infernal case has haunted me for ten days. I hereby banish it completely from my presence. To-morrow I trust that we shall hear the last of it forever."

Sharp at the hour named Inspector Stanley Hopkins appeared, and we sat down together to the excellent breakfast which Mrs. Hudson had prepared. The young detective was in high spirits at his success.

"You really think that your solution must be correct?" asked Holmes.

"I could not imagine a more complete case."

"It did not seem to me conclusive."

"You astonish me, Mr. Holmes. What more could one ask for?"

"Does your explanation cover every point?"

"Undoubtedly. I find that young Neligan arrived at the Brambletye Hotel on the very day of the crime. He came on the pretence of playing golf. His room was on the ground-floor, and he could get out when he liked. That very night he went down to Woodman's Lee, saw Peter Carey at the hut, quarrelled with him, and killed him with the harpoon. Then, horrified by what he had done, he fled out of the hut, dropping the notebook which he had brought with him in order to question Peter Carey about these different securities. You may have observed that some of them were marked with ticks, and the others – the great majority – were not.

Those which are ticked have been traced on the London market; but the others presumably were still in the possession of Carey, and young Neligan, according to his own account, was anxious to recover them in order to do the right thing by his father's creditors. After his flight he did not dare to approach the hut again for some time, but at last he forced himself to do so in order to obtain the information which he needed. Surely that is all simple and obvious?"

Holmes smiled and shook his head.

"It seems to me to have only one drawback, Hopkins, and that is that it is intrinsically impossible. Have you tried to drive a harpoon through a body? No? Tut, tut, my dear sir, you must really pay attention to these details. My friend Watson could tell you that I spent a whole morning in that exercise. It is no easy matter, and requires a strong and practised arm. But this blow was delivered with such violence that the head of the weapon sank deep into the wall. Do you imagine that this anæmic youth was capable of so frightful an assault? Is he the man who hob-nobbed in rum and water with Black Peter in the dead of the night? Was it his profile that was seen on the blind two nights before? No, no, Hopkins; it is another and a more formidable person for whom we must seek."

The detective's face had grown longer and longer during Holmes's speech. His hopes and his ambitions were all crumbling about him. But he would not abandon his position without a struggle.

"You can't deny that Neligan was present that night, Mr. Holmes. The book will prove that. I fancy that I have evidence enough to satisfy a jury, even if you are able to pick a hole in it. Besides, Mr. Holmes, I have laid my hand upon *my* man. As to this terrible person of yours, where is he?"

"I rather fancy that he is on the stair," said Holmes serenely. "I think, Watson, that you would do well to put that revolver where you can reach it." He rose and laid a written paper upon a side-table. "Now we are ready," said he.

There had been some talking in gruff voices outside, and now Mrs. Hudson opened the door to say that there were three men inquiring for Captain Basil.

"Show them in one by one," said Holmes.

The first who entered was a little Ribston-pippin of a man, with ruddy cheeks and fluffy white side-whiskers. Holmes had drawn a letter from his pocket.

"What name?" he asked.

"James Lancaster."

"I am sorry, Lancaster, but the berth is full. Here is half a sovereign for your trouble. Just step into this room and wait there for a few minutes."

The second man was a long, dried-up creature, with lank hair and sallow cheeks. His name was Hugh Pattins. He also received his dismissal, his half-sovereign, and the order to wait.

The third applicant was a man of remarkable appearance. A fierce, bull-dog face was framed in a tangle of hair and beard, and two bold dark eyes gleamed behind the cover of thick, tufted, overhung eyebrows. He saluted and stood sailor-fashion, turning his cap round in his hands.

"Your name?" asked Holmes.

"Patrick Cairns."

"Harpooner?"

"Yes, sir. Twenty-six voyages."

"Dundee, I suppose?"

"Yes, sir."

"And ready to start with an exploring ship?"

"Yes, sir."

"What wages?"

"Eight pounds a month."

"Could you start at once?"

"As soon as I get my kit."

"Have you your papers?"

"Yes, sir." He took a sheaf of worn and greasy forms from his pocket. Holmes glanced over them and returned them.

"You are just the man I want," said he. "Here's the agreement on the side-table. If you sign it the whole matter will be settled."

The seaman lurched across the room and took up the pen.

"Shall I sign here?" he asked, stooping over the table.

Holmes leaned over his shoulder and passed both hands over his neck.

"This will do," said he.

I heard a click of steel and a bellow like an enraged bull. The next instant Holmes and the seaman were rolling on the ground together. He was a man of such gigantic strength that, even with the handcuffs which Holmes had so deftly fastened upon his wrists, he would have very quickly overpowered my friend had Hopkins and I not rushed to his rescue. Only when I pressed the cold muzzle of the revolver to his temple did he at last understand that resistance was vain. We lashed his ankles with cord and rose breathless from the struggle.

"I must really apologize, Hopkins," said Sherlock Holmes. "I fear that the scrambled eggs are cold. However, you will enjoy the rest of your breakfast all the better, will you not, for the thought that you have brought your case to a triumphant conclusion?"

Stanley Hopkins was speechless with amazement.

"I don't know what to say, Mr. Holmes," he blurted out at last, with a very red face. "It seems to me that I have been making a fool of myself from the beginning. I understand now, what I should never have forgotten, that I am the pupil and you are the master. Even now I see what you have done, but I don't know how you did it, or what it signifies."

"Well, well," said Holmes good-humouredly. "We all learn by experience, and your lesson this time is that you should never lose sight of the alternative. You were so absorbed in young Neligan that you could not spare a thought to Patrick Cairns, the true murderer of Peter Carey."

The hoarse voice of the seaman broke in on our conversation.

"See here, mister," said he, "I make no complaint of being man-handled in this fashion, but I would have you call things by their right names. You say I murdered Peter Carey; I say I *killed* Peter Carey, and there's all the difference. Maybe you don't believe what I say. Maybe you think I am just slinging you a yarn."

"Not at all," said Holmes. "Let us hear what you have to say."

"It's soon told, and, by the Lord, every word of it is truth. I knew Black Peter, and when he pulled out his knife I whipped a harpoon through him sharp, for I knew that it was him or me. That's how he died. You can call it murder. Anyhow, I'd as soon die with a rope round my neck as with Black Peter's knife in my heart."

"How came you there?" asked Holmes.

"I'll tell it you from the beginning. Just sit me up a little, so as I can speak easy. It was in '83 that it happened – August of that year. Peter Carey was master of the *Sea Unicorn*, and I was spare harpooner. We were coming out of the ice-pack on our way home, with head winds and a week's southerly gale, when we picked up a little craft that had been blown north. There was one man on her – a landsman. The crew had thought she would founder and had made for the Norwegian coast in the dinghy. I guess they were all drowned. Well, we took him on board, this man, and he and the skipper had some long talks in the cabin. All the baggage we took off with him was one tin box. So far as I know the man's name was never mentioned, and on the second night he disappeared as if he had never been. It was given out that he had either thrown himself overboard or fallen overboard in the heavy weather that we were having. Only one man knew what had happened to him, and that was me, for with my own eyes I saw the skipper tip up his heels and put him over the rail in the middle watch of a dark night, two days before we sighted the Shetland lights.

"Well, I kept my knowledge to myself and waited to see what would come of it. When we got back to Scotland it was easily hushed up, and nobody asked any questions. A stranger died by accident, and it was nobody's business to inquire. Shortly after, Peter Carey gave up the sea, and it was long years before I could find where he was. I guessed that he had done the deed for the sake of what was in that tin box, and that he could afford now to pay me well for keeping my mouth shut.

"I found out where he was through a sailor man that had met him in London, and down I went to squeeze him. The first night he was reasonable enough, and was ready to give me what would make me free of the sea for life. We were to fix it all two nights later. When I came I found him three-parts drunk and in a vile temper. We sat down and we drank and we yarned about old times, but the more he drank the less I liked the look on his face. I spotted that harpoon upon the wall, and I thought I might need it before I was through. Then at last he broke out at me, spitting and cursing, with murder in his eyes and a great clasp-knife in his hand. He had not time to get it from the sheath before I had the harpoon through him. Heavens! what a yell he gave; and his face gets between me and my sleep! I stood there, with his blood splashing round me, and I waited for a bit; but all was quiet, so I took heart once more. I looked round, and there was the tin box on the shelf. I had as much right to it as Peter Carey, anyhow, so I took it with me and left the hut. Like a fool I left my baccy-pouch upon the table.

"Now I'll tell you the queerest part of the whole story. I had hardly got outside the hut when I heard someone coming, and I hid among the bushes. A man came slinking along, went into the hut, gave a cry as if he had seen a ghost, and legged it as hard as he could run until he was out of sight. Who he was or what he wanted is more than I can tell. For my part, I walked ten miles, got a train at Tunbridge Wells, and so reached London, and no one the wiser.

"Well, when I came to examine the box I found there was no money in it, and nothing but papers that I would not dare to sell. I had lost my hold on Black Peter, and was stranded in London without a shilling. There was only my trade left. I saw these advertisements about harpooners and high wages, so I went to the shipping agents, and they sent me here. That's all I know, and I say again that if I killed Black Peter the law should give me thanks, for I saved them the price of a hempen rope."

"A very clear statement," said Holmes, rising and lighting his pipe. "I think, Hopkins, that you should lose no time in conveying your prisoner to a place of safety. This room is not well adapted for a cell, and Mr. Patrick Cairns occupies too large a proportion of our carpet."

"Mr. Holmes," said Hopkins, "I do not know how to express my gratitude. Even now I do not understand how you attained this result."

"Simply by having the good fortune to get the right clue from the beginning. It is very possible if I had known about this notebook it might have led away my thoughts, as it did yours. But all I heard pointed in the one direction. The amazing strength, the skill in the use of the harpoon, the rum and water, the sealskin tobacco-pouch with the coarse tobacco – all these pointed to a seaman, and one who had been a whaler. I was convinced that the initials 'P. C.' upon the pouch were a coincidence, and not those of Peter Carey, since he seldom smoked, and no pipe was found in his cabin. You remember that I asked whether whisky and brandy were in the cabin. You said they were. How many landsmen are there who would drink rum when they could get these other spirits? Yes, I was certain it was a seaman."

"And how did you find him?"

"My dear sir, the problem had become a very simple one. If it were a seaman, it could only be a seaman who had been with him on the *Sea Unicorn*. So far as I could learn, he had sailed in no other ship. I spent three days in wiring to Dundee, and at the end of that time I had ascertained the names of the crew of the *Sea Unicorn* in 1883. When I found Patrick Cairns among the harpooners my research was nearing its end. I argued that the man was probably in London, and that he would desire to leave the country for a time. I therefore spent some days in the East End, devised an Arctic expedition, put forth tempting terms for harpooners who would serve under Captain Basil – and behold the result!"

"Wonderful!" cried Hopkins. "Wonderful!"

"You must obtain the release of young Neligan as soon as possible," said Holmes. "I confess that I think you owe him some apology. The tin box must be returned to him, but, of course, the securities which Peter Carey has sold are lost forever. There's the cab, Hopkins, and you can remove your man. If you want me for the trial, my address and that of Watson will be somewhere in Norway – I'll send particulars later."

Index

FEBRUARY.					MAY.						AUGUST.						NOVEMBER.						FEBRUARY.				
n......		7	14	21	28	Sun......	2	9	16	23	30	Sun.	1	8	15	22	29	Sun......		7	14	21	28	Sun......		6	13	20	27
on.....	1	8	15	22		Mon.....	3	10	17	24	31	Mon.....	2	9	16	23	30	Mon.....	1	8	15	22	29	Mon.....		7	14	21	28
ues. ...	2	9	16	23		Tues....	4	11	18	25		Tues.....	3	10	17	24	31	Tues....	2	9	16	23	30	Tues. ...	1	8	15	22	26
ed.....	3	10	17	24		Wed.....	5	12	19	26		Wed.....	4	11	18	25		Wed.....	3	10	17	24		Wed.....	2	9	16	23	
ur. ...	4	11	18	25		Thur. ...	6	13	20	27		Thur. ...	5	12	19	26		Thur. ...	4	11	18	25		Thur....	3	10	17	24	
i......	5	12	19	26		Fri......	7	14	21	28		Fri......	6	13	20	27		Fri......	5	12	19	26		Fri......	4	11	18	25	
t......	6	13	20	27		Sat...	1	8	15	22	29	Sat......	7	14	21	28		Sat......	6	13	20	27		Sat......	5	12	19	26	

MARCH.					JUNE.						SEPTEMBER.						DECEMBER.						MARCH.				
n......		7	14	21	28	Sun......		6	13	20	27	Sun......		5	12	19	26	Sun.		5	12	19	26	Sun......		5	12	19	26
on.....	1	8	15	22	29	Mon.....		7	14	21	28	Mon.....		6	13	20	27	Mon.....		6	13	20	27	Mon.....		6	13	20	27
ues. ...	2	9	16	23	30	Tues....	1	8	15	22	29	Tues.....		7	14	21	28	Tues. ...		7	14	21	28	Tues. ...		7	14	21	28
ed.....	3	10	17	24	31	Wed.....	2	9	16	23	30	Wed.....	1	8	15	22	29	Wed.....	1	8	15	22	29	Wed.....	1	8	15	22	29
ur.....	4	11	18	25		Thur. ...	3	10	17	24		Thur. ...	2	9	16	23	30	Thur. ...	2	9	16	23	30	Thur....	2	9	16	23	30
i......	5	12	19	26		Fri......	4	11	18	25		Fri......	3	10	17	24		Fri.	3	10	17	24	31	Fri.	3	10	17	24	31
t......	6	13	20	27		Sat......	5	12	19	26		Sat......	4	11	18	25		Sat......	4	11	18	25		Sat......	4	11	18	25	

TABLE

wing the number of days from any given day of one month, to e same day in any other month. In leap-year add 1 if February 29 in the calculation.

To the same day of	FROM ANY DAY OF											
	Jan.	Feb.	Mar.	Apr.	May	June	July	Aug.	Sept.	Oct.	Nov.	Dec.
uary..........	365	334	306	275	245	214	184	153	122	92	61	31
ruary	31	365	337	306	276	245	215	184	153	123	92	62
ch	59	28	365	334	304	273	243	212	181	151	120	90
il	90	59	31	365	335	304	274	243	212	182	151	121
....	120	89	61	30	365	334	304	273	242	212	181	151
e.........	151	120	92	61	31	365	335	304	273	243	212	182
....	181	150	122	91	61	30	365	334	303	273	242	212
ust	212	181	153	122	92	61	31	365	334	304	273	243
ember	243	212	184	153	123	92	62	31	365	335	304	274
ber	273	242	214	183	153	122	92	61	30	365	334	304
ember....	304	273	245	214	184	153	123	92	61	31	365	335
ember..	334	303	275	244	214	183	153	122	91	61	30	365

RY FOR 56 YEARS, FROM 1844 TO 1899.

EXPLANATION.

k for the year you want. Above d its Dominical letter. Look e lower part of the Diary for the letter in the line opposite the wanted. Above that letter you he days of the week, and on the e days of the month. Leap years two letters; the first serves till d of February. Leap year is the which divides evenly by 4; but ar 1900 will not be a Leap year.

	D	C	BA	G	F	E	DC
5	46	47	48	49	50	51	52
4	74	75	76	77	78	79	80
A	G	FE	D	C	B	AG	F
4	55	56	57	58	59	60	61
2	83	84	85	86	87	88	89
CB	A	G	FE	D	C	ED	C
3	64	65	66	67	68	69	70
1	92	93	94	95	96	97	98	99
5	22	29	Su	Sa	Fr	Th	W	Tu	M
6	23	30	M	Su	Sa	Fr	Th	W	Tu
7	24	31	Tu	M	Su	Sa	Fr	Th	W
			W	Tu	M	Su	Sa	Fr	Th
9	26		Th	W	Tu	M	Su	Sa	Fr

PUBLIC HOLIDAYS.

LONDON—Exchequer, and India House.—Good Friday and Christmas.

Law Offices.—Good Friday and three following days, Whit-Monday and Whit-Tuesday, Queen's Birth-day and Accession, Christmas, and three following days.

Excise, Stamp and Tax Offices.—Good Friday, Queen's Birth-day, June 28, Nov. 9, and Christmas.

Docks and Customs.—Good Friday, Queen's Birth-day, and Christmas.

IRELAND—Customs.—Good Friday, Queen's Birth-day, and Christmas.

Excise, & Stamp Offices.—Good Friday, Queen's Birth-day, June 28, Nov. 9, and Christmas.

SCOTLAND—Customs.—Good Friday, Queen's Birth-day, Sacramental Fasts, & Christmas

Excise and Stamp Offices.—Jan. 1, Good Friday, Queen's Birth-day, (Fair Saturday, Glasgow), Sacramental Fasts, and Christmas.

BANK HOLIDAYS.

ENGLAND AND IRELAND.

Good Friday, Easter Monday, the Monday in Whitsun week, the first Monday in Aug., Christmas, & the 26th of Dec. if a week day.

SCOTLAND.

New Year's Day, Christmas Day, (if either of these fall on a Sunday the following Monday shall be a Bank Holiday,) Good Friday, the first Monday of May, the first Monday of August, and any day which may be appointed by Royal Proclamation. Upon Sacramental Fast-days and other local Holidays, the Bank Offi....

POSTAGES.

INLAND LETTERS AND PACKAGES.

The rate of postage to be prepaid on inland letters and parcels of all sorts, closed or open, is as follows :—

Not exceeding 1 oz. in weight, ...1 d.
Above 1 oz. and not above 2 oz.,......1½d.
Do. 2 " do. do. 4 " ......2 d.
Do. 4 " do. do. 6 " ......2½ d.
Do. 6 " do. do. 8 " ......3 d.
Do. 8 " do. do. 10 " ......3½ d.
Do. 10 " do. do. 12 " ......4 d.

A letter above the weight of 12 oz. is liable to the charge of 1d. for every ounce, commencing with the first ounce. For instance, a letter between 12 and 13 oz. weight must be prepaid 1s. 1d. As a general rule, the postage if not paid in advance, is double the foregoing; and if the payment in advance be insufficient, double the deficiency is charged. An inland letter, for example, weighing more than an ounce, and not exceeding two ounces, and prepaid one penny only, is on delivery charged *double* the deficiency of one halfpenny, viz. one penny, and so on. —A letter directed to a person, and the person not found at the address, and the letter re-directed, single additional postage only is charged. The Post Office cannot undertake the safe transmission of valuable inclosures in unregistered letters. Letters when once posted cannot be given back upon any pretence whatever. Letters to warm climates should be gummed or wafered, not sealed with wax, as the wax is liable to get melted, to the injury of other letters.

A postmaster is not bound to re-direct letters for a person temporarily leaving his home and not having a private bag or box, unless the house be left uninhabited, or the letters would be delayed in their transmission by being sent to the house to be re-directed there. In all cases of re-direction, a written authority, duly signed by the person to whom the letters are addressed, must be sent to the postmaster.

REGISTERED LETTERS.

An inland letter or book-packet can be registered on payment of 4d. in addition to the postage of such letter or packet. The full amount in stamps for postage and registration, must be on the letter or book-packet, and it will require to be posted half an hour before the closing of the box for the mail by which it is to be despatched, otherwise a late fee of 4d. will be charged till the closing of the letter-box. A letter to be registered should be presented at the window and a receipt obtained for it, and must not be dropped into the letter-box. Any letter marked "Registered" which shall be posted without at the same time being registered shall, if observed by the post-office, be afterwards registered by the post-office, and be forwarded to its destination, charged with double the ordinary registration rate of postage. All letters containing coin are treated as registered, even though they be posted without registration, and are charged on delivery with a double registration fee, in addition to the ordinary postage. Any person sending money or jewellery in a letter, if lost or mis-delivered, has no claim on the Post-office. The most secure mode of sealing is

BOOK PACKETS,

Or plain, written, or printed paper, without a cover, or in a cover open at the ends or sides, may be sent—

Not exceeding 2 oz. in weight, ...
" 4 oz. " 1
" 6 oz. " 1
" 8 oz. " 2
" 10 oz. " 3
" 12 oz. " 3

And ½d. for every additional two ounces.

The postage to be prepaid by postage stamps affixed. If the whole postage not prepaid, the unpaid part is charged double on delivery. The packet may contain any number of separate books, printed photographs (when not on glass nor in cases containing glass), maps, parchment or vellum, either printed or written, including printed or lithographed letters, plain or mixed, but not letters, sealed or open, nor anything sealed or closed against inspection. Marking or writing allowed when no of the nature of a letter. An entry, merely stating who sends the book, or to whom is sent, is not regarded as a letter. The name and address of the sender is not only permitted but recommended, so that if the cover come off, or for any other reason the packet cannot be forwarded, it may be returned. In books and prints, all legitimate binding, mounting, or covering of the same, or of a portion thereof, will be allowed; also markers and rollers, (whether of paper or otherwise), and whatever is necessary for the safe transmission of literary or artistic matter or usually appertains thereto. For the greater security of the contents, the packet may be tied at the ends with a string. The postmaster is authorised to cut the string, but must refasten the packet afterwards. No packet can be received if it exceeds 5 pounds in weight, 18 inches in length, and 9 inches in width and 6 inches in depth, except petitions to Parliament.

INLAND PATTERN POST.

All parcels are forwarded as the sender desires, whether closed or open; the weight is limited to 12 oz., and the size to 18 inches in length by 9 inches in depth and 6 inches in width. The postage is the same as that for letters, and must be prepaid in postage stamps.

INLAND MONEY ORDERS.

For sums under 10s., - - - 1d
" 10s. and under £1, - - - 2d
" £1, " 2, - - - 3d
" 2, " 3, - - - 4d
" 3, " 4, - - - 5d
" 4, " 5, - - - 6d
" 5 " 6, - - - 7d
" 6, " 7, - - - 8d
" 7, " 8, - - - 9d
" 8, " 9, - - - 10d
" 9, " 10, - - - 11d
" 10, - - - 1s.

No order to contain a fractional part of a penny. No money order can be issued unless the applicant furnish the surname in full, and at least one initial of the Christian name of the person or firm who sends the order, and of the person or